NATURAL GAS IN NONTECHNICAL LANGUAGE

NATURAL GAS IN NONTECHNICAL LANGUAGE

Institute of Gas Technology
Rebecca L. Busby, Editor

Copyright (c) 1999 by
PennWell
1421 South Sheridan/P.O. Box 1260
Tulsa, Oklahoma 74101

Library of Congress Cataloging-in-Publication Data

Natural gas in nontechnical language / Institute of Gas Technology; Rebecca L.
    Busby, editor.
       p. cm.
       Includes bibliographical references and index.
       ISBN 0-87814-738-1
       1. Natural gas.   I. Busby, Rebecca L.   II. Institute of Gas Technology.
    TN880.N2944   1999   99-14115
    665.7--dc21   CIP

Printed in the United States of America

1 2 3 4 5 99

Table of Contents

Acknowledgments

This book could not have been written without the reliable information provided by many sources:

- The Institute of Gas Technology (IGT) provided its Home Study Course, which served as a starting point for many of these chapters. Chapter 1 of the course supplied useful background on the history of the gas industry, the origins of natural gas, and the operations of gas distribution and transmission companies.

- Additional material concerning natural gas origins and the history of the gas industry was provided by Gale Research Inc.'s *World of Scientific Discovery*, edited by Bridget Travers.

- Information on the exploration and production of natural gas came from a previous PennWell book by Norman J. Hyne, *Nontechnical Guide to Petroleum Geology, Exploration, Drilling and Production* (1995).

- Nicholas P. Biederman's series of comprehensive reports on the gas distribution industry, funded by the Gas Research Institute, provided substantial information on the construction and maintenance of distribution systems.

- The Gas Research Institute (GRI) supplied many different publications used as sources for this book, including *Gas Research Institute Digest* (GRID), Gas Appliance Technology Center (GATC) *Focus*, *GasTIPS* (Technology, Information, and Products for Supply), *GRI Baseline Projection of U.S. Energy Supply and Demand*, *Tech Profiles*, *GasFlow*, and *FrontBurner*.

GRI also provided many of the book's photos and illustrations, which were collected by Cheryl Drugan, legendary Editor of *GRID*. Other illustrations were contributed by *GATC Focus* Editor Vincent Brown and *GasTIPS* Editor Karl Lang.

My personal thanks go out to all of these sources and to Dan Handman, who patiently edited the early drafts of every chapter.

Rebecca L. Busby
San Juan Capistrano, CA
December 1998

List of Figures

List of Tables

List of Acronyms

AFV – Alternative fuel vehicle

A.G.A. – American Gas Association

APGA – American Public Gas Association

AVO – Amplitude variation with offset

CFC — Chlorofluorocarbons

CGA – Canadian Gas Association

CHP – Combined heat and power

FERC – Federal Energy Regulatory Commission

FPC – Federal Power Commission

GAMA – Gas Appliance Manufacturers Association

GRI – Gas Research Institute

GRIP – Geology related imaging program

HVAC – Heating, ventilating, and air conditioning

IAQ – Indoor air quality

IGT – Institute of Gas Technology

INGAA – Interstate Natural Gas Association of America

IPP – Independent power producer

LDC – Local distribution company

LNG – Liquefied natural gas

LPG – Liquefied petroleum gas

NGL – Natural gas liquids

NGSA – Natural Gas Supply Association

NGV – Natural gas vehicle

OPEC – Organization of Petroleum Exporting Countries

PURPA – Public Utility Regulatory Policies Act

SCADA – Supervisory control and data acquisition

SNG – Substitute natural gas

Introduction

Since its discovery thousands of years ago, natural gas has become an indispensable energy resource throughout most of the industrialized world. Many countries are fortunate enough to possess at least some domestic supplies of natural gas, while others such as Japan must import nearly all of the gas they need. Most areas that contain a wealth of oil resources are also rich in natural gas.

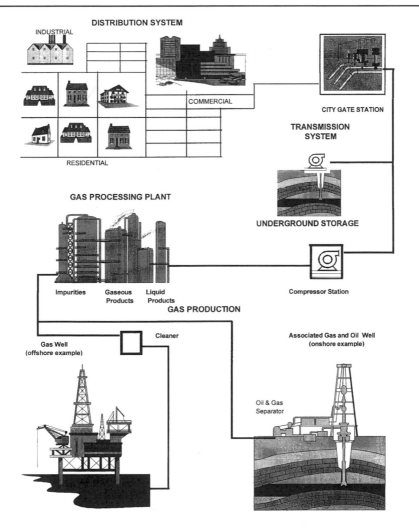

Figure I. Path of Natural Gas from the Well to the Consumer

These include Russia, the United States, the Middle East, Mexico, parts of South America, and the countries surrounding Europe's North Sea.

From a technological point of view, the major operations of any natural gas industry are exploration, production, processing, transmission (transportation by pipeline), storage, distribution, and utilization (Fig. I). These topics are discussed in the first seven chapters of this book. Even in countries that lack their own supplies, natural gas is often imported by pipeline or ship, then distributed and used in ways similar to those practiced in the U.S. and other gas-rich countries. Of course, successful operation of an industry requires more than technology and equipment. Accordingly, later chapters cover marketing and sales, governmental regulation, and future supply and demand for gas.

The U.S. natural gas industry is a large-scale operation involving production from thousands of wells and transportation of gas through hundreds of thousands of miles of large-diameter pipelines reaching into virtually every corner of the country. As one of the nation's largest industries in terms of capital investment, the natural gas industry provides about one fourth of the primary energy consumed in the United States. The strength and performance of the natural gas industry are important to the health of the U.S. economy as a whole.

CHAPTER

1

ORIGINS AND HISTORY OF NATURAL GAS

The Nature of Natural Gas

Natural gas consists mainly of methane (CH_4), the simplest hydrocarbon, along with heavier, more complex hydrocarbons such as ethane (C_2H_6), propane (C_3H_8), and butane (C_4H_{10}) (Table 1-1). The familiar gas burned in homes, businesses, and industry is virtually pure methane. The value of natural gas lies in the combustion properties of methane, a colorless, odorless gas that burns readily with a pale, slightly luminous flame. (The characteristic odor of natural gas is added artificially.)

Natural gas is the cleanest burning fossil fuel, producing mostly just water vapor and carbon dioxide. Methane is also a key raw material for making solvents and other organic chemicals. Propane and butane are usually extracted from natural gas and sold separately. Liquefied petroleum gas (LPG), which is mainly propane, is a common substitute for natural gas in rural areas not served by pipelines.

Often, natural gas also contains impurities such as carbon dioxide (acid gas), hydrogen sulfide (sour gas), and water, as well as nitrogen, helium, and other trace gases. Because carbon dioxide doesn't burn, it reduces the value of the natural gas. However, carbon dioxide can be injected into old (depleted) oil fields to enhance production, so it is sometimes recovered from natural gas and sold as

Table 1–1. Average Hydrocarbon Composition of U.S. Midcontinent Gas Production

Methane	88%
Ethane	5%
Propane	2%
Butane	1%

a by-product. Nitrogen can also be used as an oil-field injection gas, and helium is valuable in electronics manufacturing and for filling blimps and balloons.

The U.S. has a virtual monopoly on world helium production because of the large Hugoton-Panhandle gas field, which contains 0.5 to 2% helium. Nearby Amarillo, Texas, is called the helium capital of the world. This trace gas is uncommon in other natural gas reservoirs.

Hydrogen sulfide (H_2S) is very poisonous and can be lethal in very low concentrations. People can smell minute amounts of this gas, which has the foul odor of rotten eggs. "Sweet" natural gas has no detectable amounts of hydrogen sulfide. Because hydrogen sulfide is also extremely corrosive, it can damage the tubing, fittings, and valves in gas wells, so it must be removed before the natural gas can be delivered to a pipeline. In addition to removing hydrogen sulfide and carbon dioxide, most of the water is also extracted by dehydration before the gas enters the pipeline.

The Origins of Natural Gas

Generation

Nearly all natural gas is found in underground reservoirs, often associated with deposits of oil. Natural gas and crude oil were created millions of years ago by the decomposition of plants and animals that died and drifted toward the bottom of ancient lakes and oceans. Much of this organic matter was decomposed in air (oxidized) and lost to the atmosphere, but some was buried before it decayed or was deposited in stagnant, oxygen-free water, preventing oxidation.

Over the ages sand, mud, and other sediments drifted down and lithified (compacted into rock). As these layers piled up, the organic matter was preserved in the sedimentary rock. Eventually, the weight of the accumulating layers created pressure and heat that changed the organic material into gas and oil. Sedimentary "source" rocks such as coal, shale, and some limestones have a dark color that comes from their rich organic content. Many sedimentary basins are gas-prone and produce primarily natural gas.

Coal is pure woody material that has been transformed by temperature and time. Wood and coal have a specific chemistry that can generate only methane gas. That's why coal mines are dangerous and can explode. Wells are often drilled into coal beds for coal seam gas, which is pure methane formed when the woody material was changed to coal. The gas is adsorbed to the surface of the coal along natural fractures, which first produce water and then release the methane. Productive basins of coal seam gas include the San Juan in New Mexico/Colorado and the Black Warrior in Alabama.

In other sedimentary source rocks, the main factor determining whether gas or oil is formed is temperature. At relatively shallow depths, where temperatures are not high enough to generate oil, bacterial action quickly produces biogenic (or microbial) gas, which is almost pure methane (Fig. 1–1).

Commonly known as swamp or marsh gas, this biogenic gas is rarely contained; instead, it leaks into the atmosphere in enormous volumes. However, the largest gas field in the world, the Urengoy in Siberia, is biogenic in origin. The gas is trapped below the permanently frozen ground (permafrost), and the field contains

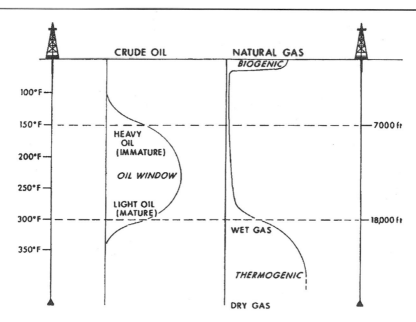

Fig. 1-1 Generation of Oil and Gas (Norman Hyne's *Nontechnical Guide to Petroleum Geology, Exploration, Drilling, and Production,* PennWell, 1995)

285 trillion cubic feet (Tcf) (8 trillion cubic meters, or m³) of gas.

At lower depths and higher temperatures (above 300°F or 150°C), thermogenic gas is created. This gas can be trapped in underground reservoirs, often beneath an impermeable "cap" rock that prevents the gas from seeping upward. In some gas reservoirs, high temperatures gasify the heavier, liquid hydrocarbons. When the gas is produced and its temperature drops, these hydrocarbons re-liquefy to form condensate. This fluid is almost pure gasoline and is often called natural gasoline. When removed with butane, propane, and ethane, the condensate is called natural gas liquids (NGL). "Wet" gas containing condensate occurs as a gas in the reservoir (even during production), but produces liquid condensate on the surface. "Dry" gas is pure methane and forms no liquid either in the reservoir or aboveground.

Higher temperatures at greater depths, below 18,000 feet (5,500 meters), also transform crude oil into natural gas and graphite (carbon), in a process similar to thermal "cracking" in a refinery, where large hydrocarbon molecules are literally broken apart. Below this depth, only gas can occur in a reservoir, and most deep wells are drilled in search of natural gas. In many cases of deep drilling into a sandstone gas reservoir, the sand grains are coated with carbon. Apparently, the original oil became too deeply buried and was thermally cracked into natural gas.

An unconventional theory of the origin of gas is the abiogenic (nonbiological) theory proposed by Thomas Gold, an astrophysicist whose ideas are viewed with skepticism by the petroleum industry. According to this theory, abiogenic gas was created when carbon, carried to the earth by meteors, bonded with the abundant hydrogen in the atmosphere to form solid hydrocarbons, which were then heated to form methane. If this theory is true, the earth could contain much more deep, primordial gas than previously thought, and in locations other than conventional sites. To test this hypothesis, a well was drilled in Sweden's Siljan Ring, site of an ancient meteor strike, but no large amounts of gas were found. Drilling was suspended in 1989 at a depth of 22,824 feet (6,957 meters), due to the difficulty of penetrating the granite rock.

Migration and accumulation

Gas can be expelled from its source rock in two ways. As the rock is buried more deeply, pressure increases and compacts the rock, thus decreasing its pore space and squeezing the gas out. Secondly, as gas is created, it increases in volume, which can fracture the rock and allow the gas to escape. Because gas is light in density, it flows upward along fractures and faults, or it can flow horizontally and then upward through permeable rock layers. This vertical and horizontal flow of gas from the source rock is called migration.

Trapping of the gas underground requires an impermeable overhead layer of cap rock. Without a trap on the migration route, the gas seeps upward, gradually flowing to the surface. In fact, most gas that has been generated over the ages has been lost rather than trapped, which is why many exploratory wells are unproductive. Also, because of migration, gas that was originally formed deeper in the earth is trapped at shallower depths.

Once the gas migrates into the trap, it goes to the top and fills the rock's pores. If oil is also present, it accumulates beneath the gas. Gas "associated" with oil occurs in contact with crude oil below ground, either in the cap of gas above the oil or actually dissolved in the oil. Associated gas contains many other hydrocarbons besides methane. "Nonassociated" natural gas does not contact oil in the trap. A nonassociated well produces almost pure methane.

The reservoir rock that holds the gas must be both porous and permeable. Porosity measures the fluid (gas or oil) storage capacity of a reservoir rock. Permeability measures how easy it is for the fluid to flow through the rock. Most reservoir rocks are permeable sandstones and carbonates, but some gas can be found in "tight" (low-permeability) formations.

Brief History of Natural Gas

The history of natural gas spans thousands of years, but as a fuel, gas did not become important to our way of life until recently, beginning in the 1930s. By the late 20th century, natural gas had become an indispensable energy resource throughout most of the industrialized world.

As early as 940 B.C., people in China piped natural gas through hollow bamboo poles to the seashore, where they used it to boil ocean water and collect salt. Some experts say that the Chinese drilled gas wells as deep as 2,000 feet (600 meters). Japanese wells were reported around 600 B.C.

Other ancient civilizations noticed the escape of natural gas from the ground and discovered that it would burn. Temples were built to house these mysterious "eternal fires," regarded by visitors with a mixture of reverence and superstition. Later reports note "pillars of fire" and a bubbling magic water that would "burn like oil." Similarly, a "burning spring" in the U.S. was described by George Washington. But these phenomena did not result in the widespread, practical use of natural gas until much later.

Birth of the industry

The emerging gas industry in America and Europe was based not on natural gas, but on "manufactured" gas, which was made by heating coal. This "coal gas" (also called "town gas"), used for lighting, transformed the way people lived during the early 1800s. Factories could operate for longer hours, and families could read newspapers and books at home after dark without expensive, hazardous candles.

William Murdoch, a Scottish inventor, was one of the first scientists to recognize that gas is a more convenient energy source than coal, primarily because it can be piped and controlled more easily. By 1792, he was lighting his own home with coal gas, despite his neighbors' fear of an explosion. Murdoch went on to develop methods of making, storing, and purifying coal gas, and the company he worked for (Boulton & Watt, of steam engine fame) began installing gas lighting in English factories. In 1802, to celebrate England's peace treaty with France, the city of Birmingam was illuminated by gas lights, initiating a whirlwind of activity for the industry.

Meanwhile, in France, Philippe Lebon experimented with the gas produced by heating sawdust, wood, and coal. In 1799, he patented a method of distilling gas from wood and invented one of the first gas lights, called the Thermolamp, which he exhibited in Paris in 1802. However, the French government rejected Lebon's ideas for a large-scale gas lighting system. Other public demonstrations of gas lights in continental Europe and England generated considerable interest.

A German entrepreneur, Frederick Winsor, came up with the idea of producing larger quantities of gas and distributing it through a centralized system. He created a joint venture, secured funding, and demonstrated gas lights for the King of England's birthday. In 1807, Winsor staged the first gas street-lighting display in London, one of the oldest such installations in the world. After a dispute with William Murdoch, Winsor was granted a charter for the first public gas distribution company, founded in 1812.

Early distribution systems used wooden pipe, which was replaced in time by metal pipe or "barrels" (made in the same way as the navy's gun barrels). Other towns and cities installed central gas plants and piping, and by 1819, London alone had nearly 300 miles (482.7 km) of gas pipes supplying more than 50,000 burners.

Across the Atlantic, American entrepreneurs kept pace with European developments. In 1802, Charles Peale tested gas lighting in his natural history museum in Philadelphia's Independence Hall. His son Rembrandt Peale was hired in 1816 to install a gas distribution and lighting system in Baltimore, where America's first gas utility was established that same year. As in England, the distribution system used wood pipe. Gas companies were soon chartered in several large Eastern U.S. cities. New Orleans saw the South's first gas company, and Montreal Canada's first.

By the late 1800s, nearly a thousand U.S. companies were selling coal gas, primarily for lighting purposes, and gas service had been extended to most major cities throughout the world. From factories and streets, gas lights moved into homes, educational buildings, and public halls, where people could debate politics well into the night.

Meeting competition

Although a meter for measuring gas had been invented in 1815, the gas consumed by most customers was not metered at first. Instead, a flat rate was charged based on the number and type of lights and their hours of use. Gas meters, which measure the volume consumed, were adopted in London in 1862, and the basic principle of early meters is still in use. In the 1890s, coin-operated meters were introduced, enabling the working class to afford gas lights and greatly increasing the number of customers.

One of the most significant inventions of this era, in 1855, was the Bunsen burner, which was capable of producing a hot blue flame. By premixing some air with the gas prior to combustion, the burner promoted more complete combustion of gas, which improved its heat release. This principle proved to be a great step forward in making gas more useful. The 1800s also saw the development of more flexible gas manufacturing processes that produced a gas with better illuminating properties.

But in the late 1800s, the fledgling gas industry was nearly killed off by the introduction of electric lighting, both the arc lamp and Thomas Edison's light bulb. Only the timely invention of the incandescent gas mantle, by Karl Auer (Baron von Welsbach) in 1885, prevented the gas lighting industry from dimming out. The cylindrical mantle (Fig. 1–2), placed over a gas flame, produced a brighter white light, while early electric lamps were relatively feeble. Even as late as 1920, a fifth of the manufactured gas being distributed was used for light-

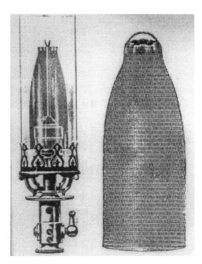

Fig. 1–2 Incandescent Gas Mantle

ing, and the mantle has since been adapted to ornamental gas lights.

Another significant advance during this time was the push-through coke oven (Fig. 1–3), which was developed to meet the large demand for blast furnace coke created by the growth of the iron and steel industries. Coke is a solid,

Fig. 1-3 Push-Through Coke Oven

porous by-product of gas manufacturing that can also be used for domestic heating. Because of coke's usefulness, many utility names retain the word "coke." By 1920, coke oven gas constituted 18.7% of all manufactured gas distributed.

The industry continued to diversify by encouraging the use of gas for purposes other than lighting. The first gas range in the U.S. was built around 1840, and in 1879, the first approach to a modern gas range, the Goodwin Company's Sun Dial Stove (Fig. 1–4), made its debut. Within four years, two other gas stove manufacturers were established, and the first distinctive gas appliance store was opened in 1887 in Providence, Rhode Island. Efforts by the gas utility companies to promote gas for cooking were so successful that by 1900 it was the industry's most important use.

These promotional efforts also resulted in a significant increase in the use of gas for water heating. Gas burners were applied to water storage tanks in the early 1860s. The circulating water heater, a cheap and effective device, first

Fig. 1-4 Goodwin Cooking Stove

appeared in 1883. A water heater with thermostatic controls, forerunner of automatic water heaters, was developed a few years later.

Transition from manufactured to natural gas

In the early 1800s, wells drilled for water and brine often produced natural gas accidentally. For the most part, this gas was regarded as a nuisance because it interfered with the intended use of the well. Infrequently, attempts were made to use the gas on a small scale. In 1821 in Fredonia, New York, a gunsmith, William Hart, drilled America's first natural gas well, which was covered with a large barrel. The gas from this shallow well (27 feet, or 8.2 meters deep) was piped through wood to nearby homes. A few years later, the natural gas was used to light the town's streets in honor of General Lafayette's visit.

Through the 1830s and 1840s, a few other gas wells were drilled in Pennsylvania, New York, and West Virginia, including a 1,000 foot (300 meter) deep well near George Washington's "burning spring". The natural gas in this well had sufficient pressure to send a column of water 150 feet (50 meters) into the air. During this period, natural gas was used only in places that happened to be near gas wells, mainly because early pipes were unable to transport it long distances.

By the time the first natural gas company was established in Fredonia in 1865, oil had been discovered near Titusville, Pennsylvania, where the first successful oil well was drilled. In the oil rush that followed, drillers shunned natural gas. It could not be captured and used without installing a pipeline, whereas oil could be carted away to centers of usage. Gas wells discovered in the search for oil were not controlled, but instead were allowed to flow for weeks or months in the hope that oil would ultimately appear. Gas produced along with oil was usually just burned off, or flared.

Natural gas was first used in the iron and steel industry in Pennsylvania, which stimulated this application for gas, particularly in the Pittsburgh area. Several companies were organized to transport gas over short distances to steel mills throughout the state. In 1885, Andrew Carnegie noted that natural gas used for steelmaking had replaced 10,000 tons (9,070 metric tons) of coal a day. This boom was short-lived, however, because the known local gas reservoirs were soon depleted. By 1900, steel mills in the Pittsburgh area were returning to coal.

For the next 25 years, natural gas supply repeated this boom-and-bust pattern. Wastefulness contributed to rapid depletion of some early gas fields, and leakage from poorly constructed pipelines was excessive. By 1920, the worst of these excesses had been curbed, but large amounts of gas continued to be flared in some oil fields into the early 1950s.

Pipeline network expands

Gradually, as pipeline technology improved and natural gas was discovered in larger quantities, the industry began to grow again. During oil drilling operations in West Bloomfield, New York, in 1870, a very high, open flow of natural gas was discovered. Although wasted at first, the gas was later transmitted to Rochester through the first "long-distance" pipeline (Fig. 1–5), which was made of pine logs and was only about 25 miles (40 kilometers) long and less than a foot (30 cm) in diameter.

High-pressure transmission of natural gas was initiated in 1891 by the Indiana Gas and Oil Company, which laid two parallel pipelines made of wrought iron over a distance of 120 miles (198 km) from an Indiana gas field to Chicago. The pipelines were operated at a pressure of 525 psi (3,620 kPa).

Uncontrolled exploitation and waste exhausted this field by 1907, and manufactured gas was piped instead.

Fig. 1-5 Wood Pipe Used in Early Line to Rochester

When huge amounts of natural gas were discovered in Texas, Oklahoma, and Louisiana, pipelines were installed to supply markets relatively close to the fields. Long-distance transportation of gas still was not considered economically or technically feasible. Also, the transportation of this gas to the North and East was opposed by railroads and others who would lose markets for competing fuels.

But when seamless steel pipe was introduced in the 1920s, natural gas transportation began to make a profit. The strength of this pipe allowed transmission of gas at higher pressures and thus in greater quantities. Oxyacetylene welding had been introduced in 1911 as a means of joining steel pipe into long sections, and methods for measuring large quantities of gas were developed. High-strength electrically welded pipe was introduced by A. O. Smith in 1927.

Long-distance transmission of natural gas lowered its cost and made it more competitive with other fuels, leading to the extensive use of gas for space heating.

This gas load, in turn, justified construction of more transmission lines, including parallel lines to provide greater supply to cities already using gas. To avoid early gas field depletion and match gas supply with demand, integrated companies were formed to handle production, transmission, and distribution of natural gas from the wellhead to the customer. Before 1925, the longest lines were about 300 miles (500 km) in length, but by 1931, several long-distance transmission systems had been constructed.

Pipeline construction paused somewhat during the 1930s and 1940s, due to the Depression of 1929 and restrictions on the use of steel during World War II. During this period, the U.S. gas industry continued to be based almost entirely on the distribution of manufactured gas, except for areas with easy access to large natural gas fields.

Very little gas-making equipment was built during the Depression, so the war's tremendous demand for fuel caught most manufactured gas companies desperately short. Plants once condemned as obsolete were reactivated. But as soon as World War II ended, the economy revived and construction of long-distance gas transmission lines boomed. By 1950, gas pipelines eclipsed those used to transport oil.

Bibliography

"The Gas Range and How It Grew," *Control Tower* 2, 6-7 (1960) First Quarter.

Harper, R.B., "Outline History and Development of the Gas Industry." Unpublished notes by the author, 1942. IGT Technology Information Center.

Hilt, L., "Chronology of the National Gas Industry," *American Gas Journal* 172, 29-36 (1950) May.

"How Man Made a Substitute for the Sun," *Baltimore Gas and Electric News* 5, 272-35 (1916) June.

Hunt, C., "Gas Lighting," Vol. III, 232, 300, in *Chemical Technology*, Groves, C. E., and Thorp, W., Eds.

Hunt, C., A *History of the Introduction of Gas Lighting*. London: W. King, 1907.

Hyne, Norman J., *Nontechnical Guide to Petroleum Geology, Exploration, Drilling and Production*. Tulsa, OK: PennWell Publishing Company, 1995.

Norman, O.E., *The Romance of the Gas Industry*. Chicago: A.C. McClurg and Co., 1922.

"Steps Taken in Third Era to Eliminate Gas Waste," *Oil & Gas Journal* 33, 115-16 (1934) August 27.

"The Story of Gas," (12-part series), *A.G.A. Monthly*, July/August 1975 ff.

Stotz, L., and Jamison, A., *History of the Gas Industry*. New York: Stettinger Bros., 1938.

Suttle, R.R., "Chronology of the Southern Gas Industry 1802-1957," *American Gas Journal* 178, 29-33 (1953) May.

Travers, Bridget, Ed., *World of Scientific Discovery*. Detroit: Gale Research Inc., 1994.

2 EXPLORATION PRINCIPLES, TOOLS, AND TECHNIQUES

How Natural Gas Is Trapped Underground

Most of the world's resources of petroleum (crude oil and natural gas) lie deep underground in traps of one kind or another. Over the ages, layers of rock are subjected to intense pressure that causes them to rise, sink, or shift from side to side. Rocks are also deformed by weathering and erosional processes that transport and deposit sediments. During all these geological changes, natural gas can become trapped underground in reservoirs, or isolated zones that produce gas or oil. Reservoir-quality rock must be sufficiently "sponge-like" (porous) to hold hydrocarbons, and the rock's pores must be interconnected (permeable) to allow gas flow.

In addition to reservoir-quality rock, the accumulation of natural gas and crude oil depends on a source of organic matter to generate hydrocarbons, pathways for the hydrocarbons to move into the reservoir rock, and adjacent rock that traps the hydrocarbons and prevents them from migrating further. Natural gas in a single reservoir has specific characteristics, but these can differ greatly from one reservoir to another in the same gas field.

Structural Traps

Structural traps are created when pressure and other geological processes deform or fracture the reservoir rock. Most often, gas and oil are found in geological structures called anticlines, where layers of rock have been gently folded upward to form an arch above the petroleum. If the rock layers are folded downward instead of upward, the structure is called a syncline. Domes are uplifts similar to anticlines. They both form high points in the reservoir rock. These structures were the first type of hydrocarbon traps to be recognized by exploration geologists. Domes and anticlines are usually asymmetrical and contain more than one gas-producing layer.

When rocks are fractured and large sections have slipped past one another, the structure is called a fault. Faults are classified by whether the rock sections shifted up and down (dip-slip) or sideways (strike-slip). Figure 2–1 illustrates the various types of petroleum traps.

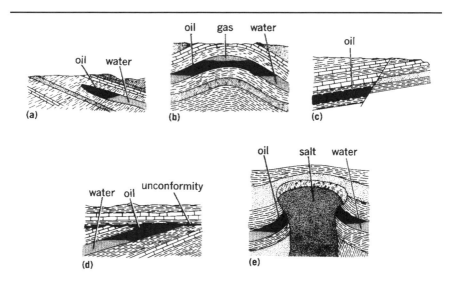

Fig. 2-1 Geological Traps Containing Gas and Oil

Sometimes, a fault cuts through a trap and divides it into numerous smaller compartments sealed off from one another. Thrust faults, which are caused by compression, reveal themselves on the surface as mountain ranges. In these structures, one piece of rock has been thrust up and hangs over the lower piece of rock. The Rocky Mountain range is a zone of thrust faults called an overthrust belt.

Fractures can also trap gas by increasing the permeability of fine-grained sedimentary rocks such as shales and chalks. Although these rocks are porous, they lack the permeability that allows gas to flow into them and become trapped. When fractures make them permeable, these sedimentary rocks can become reservoir rocks.

Stratigraphic traps

In addition to structural traps, petroleum is found in stratigraphic traps, which capture gas and oil within the strata, or layers, of rock. Stratigraphic traps are formed when the rock's porosity or permeability changes, preventing gas from migrating out of the rock. In general, stratigraphic traps are harder to discover than structural traps.

Stratigraphic traps are created while the reservoir rock is being deposited – for example, during the accumulation of sandstone in a river bed or the growth of limestone into an underwater reef. A sequence of geological processes – gradual deposition, shifting, exposure to weather, erosion, and reburial of sedimentary rock – results in "angular unconformities" (Fig. 2–2), which can trap huge amounts of gas and oil if overlain by an impermeable cap.

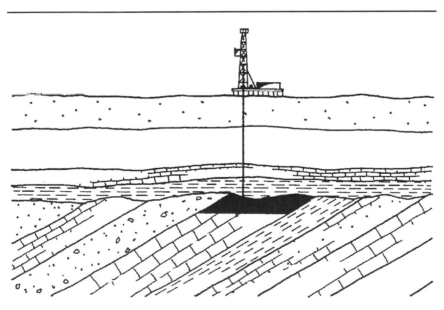

Fig. 2–2 Angular Unconformity Trap (Norman Hyne's *Nontechnical Guide to Petroleum Geology, Exploration, Drilling, and Production*, PennWell, 1995)

Combination traps

Combination traps have both structural and stratigraphic elements. The largest gas field on the North American continent – the Hugoton-Panhandle field stretching across Texas, Oklahoma, and Kansas – is a combination trap that will eventually produce 70 Tcf (2 trillion m³) of natural gas. This field covers an enormous area that is 275 miles (443 km) long and from 8 to 57 miles (13 to 92 km) wide.

A salt dome is another type of combination trap where a large mass of salt rises through sedimentary layers above it to form a plug-shaped structure. Hundreds of these salt domes lie in sedimentary rocks on the bottom of the Gulf of Mexico and along the Gulf's coastal plain.

Huge gas volumes can also be contained in carbonate rock reservoirs with a very complex geology. These reservoirs were formed when ancient caves were created on the surface by water dissolving the rock. The caves were gradually buried and eventually collapsed, sending out fractures and creating a "paleocave system." These systems can form compartments or coalesce into larger zones connected by fractures, forming a complex, heterogeneous reservoir up to several thousand yards across.

Plays and trends

When geologists have shown that an area contains commercial quantities of natural gas or crude oil, they call it a "play." The play consists of a proven combination of reservoir rock, trap, and cap rock or other seal. A "trend" is the fairway along which the play has been proven and where more fields could be found (Fig. 2–3).

A prospect is the exact location where the geological and economic conditions are favorable for drilling an exploratory well. Four geological factors determine the success of a particular prospect: source rock that has generated gas or oil, reservoir rock to hold the gas, a trap to seal it off, and the right timing. The trap has to be in place before the gas migrates away from the area.

Exploration Techniques

In the early days of the petroleum industry, the explorer's tools were as simple as a pick and shovel, and deposits of crude oil and natural gas were often discovered by accident. Exploratory wells were drilled randomly or, with greater success, located near known seeps of gas above ground.

More recently, however, people exploring for natural gas use a variety of sophisticated methods classified as geological, geochemical, or geophysical techniques. These tools are used to search for geologic traps that seal hydrocarbons in porous, permeable reservoirs. In general, these reservoir rocks are sedimentary, mainly sandstones and carbonates, deposited in ancient lakes, rivers, and oceans.

Fig. 2-3 Tuscaloosa Trend Containing Deep, Wet Gas (Norman Hyne's *Nontechnical Guide to Petroleum Geology, Exploration, Drilling, and Production*, PennWell, 1995)

Geological methods

Geological exploration techniques involve drawing maps of surface and subsurface structures and taking samples of rock formations. Topographical maps, for example, show the elevation of the earth's surface, with curvy contour lines indicating areas of equal elevation.

Some geologic maps show where layers of rock "outcrop," or expose themselves on the surface. These maps are flat, two-dimensional representations of the earth's surface. The third dimension – that is, the orientation of the rock layers – is shown by a symbol called strike-and-dip, which indicates the horizontal and vertical orientations of the plane.

The basic rock layer used for geologic mapping is the formation, which has a definite top and bottom. All rocks on the earth's crust have been classified into formations by geologists. Each formation has a two-part name: 1) its geographic location, usually a small town, and 2) its dominant rock type, such as sandstone or limestone. To show the vertical sequence of these rock layers, geologists use the stratigraphic column, where the youngest formation appears on top and the oldest at the bottom (Fig. 2–4).

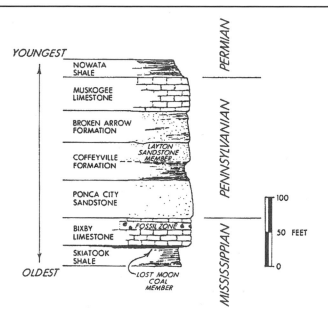

Fig. 2–4 A Stratigraphic Column (Norman Hyne's *Nontechnical Guide to Petroleum Geology, Exploration, Drilling, and Production,* PennWell, 1995)

Besides mapping structures on or near the surface, geologists use contour lines to illustrate structures below ground. These subsurface maps include three important types:

- Structural, which shows the elevation of rock layers
- Isopach, which shows their thickness
- Lithofacies, which shows the variations (facies) in a single rock layer

In addition to mapping, geologists extract cores from underground rock formations and collect "cuttings," or chips of rock while drilling exploratory wells. Cores and cuttings provide extremely useful information for assessing the formation's lithology, hydrocarbon content, and ability to hold and produce gas.

Geologists use these samples to determine the conditions under which the rock layers were formed and to evaluate whether these conditions were favorable for the generation, accumulation, and trapping of hydrocarbons.

A lot of subsurface information from existing wells, no matter who has drilled them, eventually becomes public. Government regulations require that certain well "logs," or records, be released within a specific time (from months to years, depending on the area). This geological information might come from drilling for other resources such as water, brine, coal, and minerals.

Another important source of geological information is the regional well log library. Almost every drilling area in the country has a well log library, but membership costs money. Similarly, regional libraries collect other drilling information such as well cuttings and cores. Often, these can be examined for a fee.

To help exploration geologists, the natural gas industry has compiled information on undeveloped gas resources. Atlases of gas basins have been published to reduce or eliminate the process of mapping gas reservoirs. These atlases and accompanying databases cover gas plays in six areas: the Rocky Mountains, the Appalachians, Texas, the Gulf of Mexico, the Central and Eastern Gulf Coast, and the Mid-Continent. The data compiled in these publications helps identify areas with the highest exploration potential and concentration of unrecovered hydrocarbons in existing fields. The atlases also help reveal the most prolific combinations of structures and rock type.

Geochemical methods

Geochemical techniques are based on analyzing the chemical and bacterial composition of soil and water on the surface, above or near underground gas or oil reservoirs. This composition is often altered by the very slow escape of hydrocarbons to the surface. Thus analysis of soil or water samples from an area could suggest the presence of petroleum in the rocks below, if traces of hydrocarbons are found. In many cases, minute seeps of petroleum on the surface are invisible to the naked eye. These "micro-seeps" often occur in a pattern called a hydrocarbon halo.

One geochemical method used to determine the maturity of a source rock is vitrinite reflectance. Vitrinite is a type of plant organic matter found in shale. The source rock sample is polished and then examined under a reflectance microscope. The percentage of light reflected from the vitrinite depends on the maturity of the source rock. Vitrinite reflectance can indicate whether gas and oil have been generated.

Geophysical methods

The science of geophysical exploration was pioneered by the American

Everette DeGolyer, who detected the first oil-bearing salt dome in 1924. He also developed the geophysical technique called seismic reflection, which is still one of the exploration geologist's most important tools. Seismic techniques and other geophysical methods enable geologists to determine the depth, thickness, and structure of subsurface rock layers and evaluate whether they are capable of trapping natural gas and crude oil. In effect, seismic technology lets geologists "see" beneath the surface of the earth (Fig. 2–5).

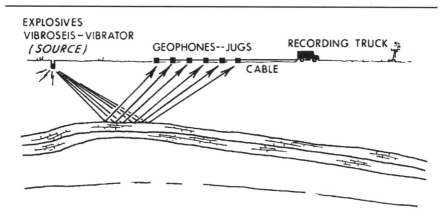

Fig. 2–5 The Seismic Method on Land (Norman Hyne's *Nontechnical Guide to Petroleum Geology, Exploration, Drilling, and Production*, PennWell, 1995)

Seismic reflection reveals underground structures by bombarding them with strong sound waves (shock waves or vibrations) and then measuring how the waves are reflected back upward by the underground rock. Seismic methods help discover hydrocarbon traps by creating an image of the subsurface rock layers. The greatest advances in petroleum exploration during the past several decades have involved innovative ways to acquire seismic data and analyze it using computers. These improvements are providing sharper images of underground structures at greater and greater depths.

Seismic equipment used in petroleum exploration works like a depth sounder on a boat, which emits periodic bursts of sound that bounce off the ocean floor. Depth is indicated by the time it takes for the sound waves to travel down and back, and different materials reflect the waves at different velocities. A detector on the surface records the signals from underground, along with unwanted noise that must be filtered out.

On land, the most common methods of generating seismic energy are explosives and vibration. Dynamite was the first seismic energy source, and other materials have been used such as primacord (a length of explosive cord). However, explosives are dangerous and expensive, which is why Conoco developed Vibroseis™. This technique uses a vibrator truck equipped with hydraulic motors mounted on the truck bed. A pad is lowered from the truck to the ground until most of the truck's weight rests on the pad. Then the hydraulic motors use this heavy load to shake the ground. Vibration equipment is convenient because it is portable and can be used in populated areas.

Seismic energy travels down through the rock, and each time the sound wave strikes the top of a rock layer, part of the energy is reflected back to the surface as an echo. The rest of the energy continues to travel downward and bounce off deeper layers or dissipate. Returning echoes are detected on land by vibration sensors called geophones ("jugs"), which translate the signals into electrical voltage. The locations of the seismic source and geophones are pinpointed by a survey crew.

Seismic operations at sea are based on the same principles as land methods but use some different tools. Seismic energy is usually generated by an air gun that emits high-pressure bubbles into the water. Surveying is done by Loran radio transmitters or global positioning satellites. The seismic source is towed by a boat, and the vibration detectors are called hydrophones.

The results of seismic tests are recorded and used to analyze subsurface rock layers. Any deformation of the rock, such as tilting, faulting, or folding, is apparent on a seismic record. "Bright spots" often successfully reveal the location of natural gas reservoirs and caps of gas lying over oil fields. The bright spot is an area of intense reflection on the seismic profile (Fig. 2–6). It is produced by an echo of about 20% of the seismic energy. However, not all bright spots are commercial gas deposits. Another indicator is a flat spot, which is a reflection off a gas-oil or gas-water contact.

Recent advances in seismic technology include AVO (amplitude variation with offset), which enhances analysis of bright spots in surface seismic data. Another approach is called GRIP (geology related imaging program). Subsurface geology often exhibits structural and lithologic complexities that mask the exploration target. The GRIP method increases the resolution of the seismic image by incorporating geologic information directly into the seismic survey design.

In addition to two-dimensional seismic imaging methods, geologists can employ 3-D techniques, which give a three-dimensional, highly detailed seismic image of the subsurface. This method is similar to the use of CAT scans (computer aided tomography) and MRI (magnetic resonance imaging) for obtaining images inside the human body. Three-dimensional techniques are continually being improved as geologists learn how to correct for noise (unwanted signals) caused by variations in the compaction and thickness of rock layers.

Another recent advance is "cross-well" seismic technology, which uses a seismic energy source in one well and signal receivers in one or more nearby wells. The data images can then be used to characterize the rock formations between the wells. Cross-well seismic images have much greater resolution than surface

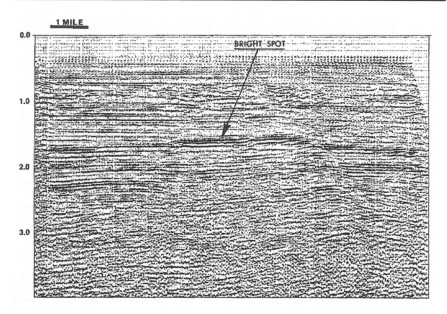

Fig. 2–6 Seismic Record in the Gulf Coast over a Gas Field, Acquired by Vibroseis™ (courtesy of Conoco)

seismic data. Cross-well imaging has proven successful in oil exploration, but has only recently been applied to natural gas fields.

In addition to seismic techniques, geophysical methods include measurements of gravity and magnetic force. The geologist's gravity meter reveals variations in gravitational force, while the magnetometer measures the strength and direction of the earth's magnetic field.

Gravity meters are very sensitive to the density of the rocks in the subsurface. They can detect a mass of relatively light rock, such as a salt dome or porous reef, or a mass of heavier rocks such as the core of a dome or anticline. Magnetometers are used primarily to detect variations in the elevation of the basement rock. Magnetic data can be used to estimate the thickness of sedimentary rocks filling a basin and to locate faults that displace basement rock.

Modeling

In cases where all these techniques have had limited success, exploration geologists turn to mathematical models to "imagine" the underground geological structures and depositional conditions. To build these computer models accurately, explorationists combine well data and geophysical information into a hypothetical picture of the subsurface. Recent advances have greatly increased the success of computer modeling.

Exploratory wells

Geological, geochemical, and geophysical techniques, no matter how sophisticated, can only indicate where natural gas or crude oil deposits *might* be found. The only method of proving the actual existence of gas or oil underground is to drill a well and test the contents of the target formation.

Because drilling is very costly, all the information obtained by the various exploratory techniques is usually correlated to select the most promising site. Wells drilled to discover a new gas or oil field are called exploratory or "wildcat" wells. The cost of a single wildcat well can range from $100,000 to more than $15 million. In spite of all the geologist's preliminary work, only about 15% to 20% of these wells discover gas or oil in amounts that are economical to produce.

If a wildcat well does discover a new gas field, it is called the discovery well for that field. The ultimate size of the field is determined by drilling more "step-out" wells surrounding the discovery well. Once the gas field is defined, additional wells develop the field or increase its gas production rate (in-fill wells). The methods used by drillers to evaluate the well's productivity are discussed in Chapter 3, *Drilling, Production, and Processing.*

Recent Targets for Natural Gas Exploration

In the early 1970s, the industrial world received an economic shock when the Organization of Petroleum Exporting Countries (OPEC) virtually turned off the oil spigot, causing energy prices to skyrocket. This act ushered in the energy crisis and forced industrial countries to reduce their consumption of oil, gas, and other energy resources. People's lifestyles changed dramatically. To use less gasoline, cars became smaller and lighter. To conserve heating and cooling energy, buildings were constructed with less ventilation. Nearly all industrial processes were modified to become more efficient.

As a result of this economic transformation, many marginal gas and oil reserves suddenly became profitable to produce. Exploration and production technologies were developed to exploit these resources and increase the world's supply of energy. Although energy prices have long since returned to more moderate levels, these technologies are still useful and are improving all the time. Natural gas reserves that had been considered uneconomic included gas shales, "tight" (low-permeability) gas sandstones, and coal seam gas (coalbed methane).

Organic-rich shales cover a large portion of the eastern and central regions of the U.S. Exploration targets during the 1970s and 1980s focused on the Antrim gas shale in the Mississippi and Michigan basins. Gas Research Institute estimates that exploration research on gas shales has increased the amount of proven gas reserves by almost 2 Tcf (60 billion m³).

Similarly, exploration of tight gas sandstones has identified one of America's largest concentrations of undeveloped natural gas, which lies in four basins of tight, stacked sandstones in the western U.S. The primary exploration target has been the Greater Green River basin in Wyoming and Colorado. Research has concentrated on how to increase the permeability of these sandstones.

Coal seam gas is pure methane created when woody material is transformed to coal. Traditionally, coal seam gas was regarded as a curiosity, and no one could figure out how to recover the gas economically. Since the 1970s, geologists have learned to understand the fundamentals of producing this coalbed gas, making it a profitable exploration target. Natural gas production from coalbed reservoirs has grown from virtually nothing in the 1980s to nearly 1 Tcf/yr (30 billion m³/yr), representing more than 5% of annual U.S. gas production.

Other natural gas resources emerging during the 1990s include gas from very deep wells, 15,000 feet (5,000 meters) or more below the ground. This gas is hard for geologists to find and develop because of the extreme pressures and temperatures that occur that deep in the earth.

Bibliography

Cicchetti, E., "The Quest in Exploration Research: To Reduce the Cost of Finding Natural Gas," *Gas Research Institute Digest* 19, Winter 1997/1998 (No. 4), pp. 18-21.

Hyne, Norman J., *Nontechnical Guide to Petroleum Geology, Exploration, Drilling and Production.* Tulsa, OK: PennWell Publishing Company, 1995.

Travers, Bridget, Ed., *World of Scientific Discovery*. Detroit: Gale Research Inc., 1994.

CHAPTER

3

DRILLING, PRODUCTION, AND PROCESSING

Basic Steps of Drilling and Production

D rilling for natural gas begins after a prospect has been identified as having geological and economic conditions favorable for gas production. After getting legal permission to drill, operators begin by spudding the well, or breaking ground. Once the hole is drilled, measurements called well logs are performed to identify the gas-bearing formations. These tests also evaluate the porosity and permeability of the rock.

If the well looks promising, the bottom of the bare hole is cased, or lined with metal pipe to seal it from the rock. Completion of the well also involves setting its foundation underground with cement or other materials. Then holes are shot through the casing and cement to allow gas to flow into the well. Finally, smaller diameter tubing is run down the hole to conduct the gas to the surface.

For gas wells that target formations having a low permeability, extra steps might be required, such as fracturing. Also, different types of drill rigs and procedures are needed for offshore wells, especially those in deeper waters.

Drilling Mechanics

A team of geologists, geophysicists, and engineers selects the well site and its drilling target, or potential reservoir rock. They also estimate how deep the well should be drilled to exploit that target. The average gas well in the U.S. is drilled to about 5,800 feet (1,800 meters). However, drilling depths vary greatly from well to well depending on location, rock properties, and other factors.

Preliminaries

In the U.S., drilling operators must identify the owner of the land's mineral rights, as opposed to the surface rights owner, which can be a different person. About one-third of the mineral rights on land in the U.S. is owned by federal or state governments. In other countries, the federal government owns the land's mineral rights. Operators must get a lease signed giving them permission to explore, drill, and produce gas for a given period of time. The mineral rights owner receives a royalty based on a percentage of revenue generated by the well, free and clear of production costs.

In offshore areas of the U.S., states own the mineral rights out to three nautical miles from shore, while the federal government owns sea-bottom rights on the outer continental shelf (to a depth of about 8,000 feet, or 2,440 meters, of water). In foreign waters such as the North Sea, mineral rights have been divided among the countries surrounding the area.

Drilling rigs are usually owned and operated by contractors, who agree to drill the well to a specific depth and target. This drilling contract also includes equipment specifications and a spud date for starting the well. After a surveyor pinpoints the site, a large pit is dug and lined with plastic. This pit holds unneeded drilling mud (lubricating fluids), cuttings, and other materials from the well. If the well is shallow, the entire drilling rig comes on the back of a truck or trailer. Otherwise, several modules are transported to the site.

Cable-tool drilling

Cable-tool rigs have been used for centuries to drill for fresh water or for brines that were evaporated for salt. The cable-tool process is relatively simple. The rock is fractured by pounding it with a chisel-like drill bit, which is suspended by a cable from a tower, or derrick. An engine moves a wooden walking beam up and down, which raises and drops the bit. Cable-tool drilling is very slow, and the hole quickly becomes clogged with rock chips, requiring the bit to be raised while the well is cased so that drilling can resume. Because of these drawbacks, cable-tool rigs are no longer used extensively for drilling gas wells.

Rotary drilling

Rotary drilling techniques were introduced throughout the world between 1895 and 1930. By 1950, about half of all rigs were the rotary type, and since then, almost all wells have begun to be drilled with rotary equipment. Its biggest advantage compared to cable-tool drilling is speed – it can drill several hundreds or thousands of feet in one day.

In rotary drilling, the rig spins a long length of steel pipe with a bit mounted on the end. Rotation of the bit digs into the earth and creates the wellbore, or borehole. The drill rig is powered by a diesel engine or other prime mover with a capacity of about 1,000-3,000 horsepower (750-2,200 kilowatts). The engine turns a circular rotary table on the floor of the derrick.

The rig's rotating system (Fig. 3–1) includes the drillstring (turning drillpipe, bit, and related equipment) and the "kelly," a piece of square pipe that is gripped and turned by the rotary table. The drillpipe is made of heat-treated alloy steel, typically in sections of 30 feet (9 meters). Sections of drillpipe are threaded together as they go downhole.

Most drill bits are shaped like three cones melded together, with teeth along

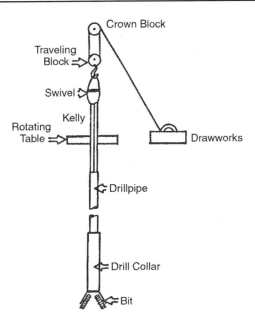

Fig. 3–1 **Rotating System of a Rotary Drill Rig (Norman Hyne's** *Nontechnical Guide to Petroleum Geology, Exploration, Drilling, and Production,* **PennWell, 1995)**

the cone edges (Fig. 3–2). Bits turn slightly faster at shallower depths and slow down as they dig deeper. Some of the drillstring's weight rests on the bit, exerting pressures of about 3,000 to 10,000 psi per inch of bit diameter (8,144 to 27,142 kPa per centimeter). The average bit wears down after about 40 to 60 hours of digging. To change the bit ("make a trip"), the drillstring must be pulled out of the well.

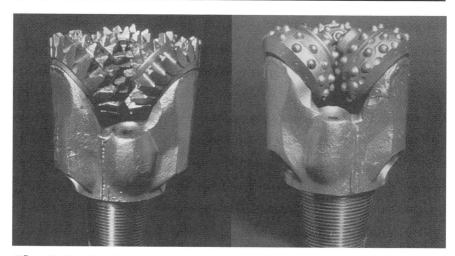

Fig. 3–2 Milled-teeth Tricone Drill Bit (Norman Hyne's *Nontechnical Guide to Petroleum Geology, Exploration, Drilling, and Production,* PennWell, 1995)

To clean and lubricate the rotating bit as it cuts downward through the rock, drilling mud is circulated in the hole. This mud is a mixture of clay, weighting materials, chemical additives, and water or oil. The drilling mud is pumped down the pipe, through holes in the bit, and back up to the surface through the annular space between the rotating drillstring and the borehole walls. Cuttings from underground formations are carried up by the mud.

Besides lubricating the bit, the drilling mud plasters the wall of the hole with solid clay particles (filter cakes), which help support loose formations and seal off water-bearing layers of rock. The density, viscosity, and other properties of the drilling mud are frequently checked. Sometimes, a drilling problem requires a mass of chemical additives ("pill") to be injected into the mud.

The tool pusher is the drilling company employee who is similar in authority to the captain of a ship. The tool pusher supervises all drilling operations and usually lives at the drillsite full-time. Other key players are the drilling fluids engineer ("mud man," often representing a petroleum field service company), the driller in charge of each crew shift (tour), the derrickman, and on offshore rigs, the roustabouts, who handle equipment and supplies carried on and off the rig.

Common drilling problems

Aside from the risk of drilling a dry hole with no gas, one of the most common problems is that something breaks inside the well, such as a piece of bit or drillstring, or falls down into it, such as wrenches or other tools. These pieces of metal are called fish or junk. Because the bit can't dig through them, drilling must be suspended until they can be retrieved by "fishing" with special tools leased from a service company.

Other drilling problems result from high pressures underground, which get higher as the well goes deeper. During drilling, the pressure of the mud should be slightly higher than the hydrostatic pressure of the overlying water. This creates an overbalance, preventing fluids from entering the well prematurely.

When unexpectedly high subsurface pressures are encountered, gas or water can flow into the well, dilute the drilling mud, and reduce its pressure. This is called a kick, and it can lead to a worse situation called a blowout, or uncontrolled flow of fluid. To avoid these events, blowout preventers are used to close off the top of the well.

Kicks and blowouts can be detected aboveground in many ways, such as monitoring the drilling mud. When a kick is detected, the well is closed by the blowout preventers, and heavier mud ("kill mud") is pumped down to circulate the kick out of the well. Blowout preventer tests are run periodically on rigs to test the equipment and the drilling crew's reaction time.

Drilling techniques

To exploit a gas or oil field efficiently, governments often regulate well spacing. Only one gas well can be drilled and completed over a given area, typically 640 acres (2.6 million m^2). In the U.S. and Canada, the amount of gas production from a well, lease, or field is also limited over a specific unit of time.

Directional Drilling. Traditionally, most wells have been drilled as straight holes, with little deviation from the vertical. But more recent rotary rigs allow directional (deviated) drilling to reach a specific target that is not accessible by a straight hole (Fig. 3–3). For example, a well can reach a target that lies underneath a densely populated area by drilling from a surface location outside the area. Directional drilling is also used to tap multiple producing zones, drill around a fish in the hole, or reach offshore reservoirs from land, where drilling is cheaper. Also, many offshore rigs save a lot of time and money by drilling multiple directional wells from a single floating platform.

In directional drilling, the spot where the hole deviates from the vertical is called the kickoff. This point is followed by a curved portion called the dogleg or build angle. A recent development that is very useful for directional wells is turbodrilling, where the bit is rotated by a downhole turbine powered by the circulating mud. Because the rotary motion is imparted only at the bit, rotation of the drillstring is unnecessary.

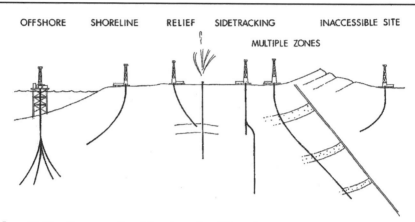

Fig. 3–3 **Reasons for Directional Drilling (Norman Hyne's *Nontechnical Guide to Petroleum Geology, Exploration, Drilling, and Production,* PennWell, 1995)**

Horizontal Drilling. The directional drilling concept has been extended to drilling wells horizontally, and this practice has become more important to gas and oil exploitation. As opposed to customary wells, where only the thickness of the reservoir is exposed for gas production, a horizontal well running parallel with the formation exposes a much greater length of reservoir rock.

Like the directional well, a horizontal well has a kickoff point where the wellbore begins to build an angle. But the angle continues to increase until the wellbore is penetrating the formation laterally. Horizontal wells might be drilled in order to:

- Increase recovery from a thin formation
- Increase gas production from a low-permeability reservoir
- Intersect isolated productive zones
- Improve gas recovery by connecting vertical fractures
- Prevent production of excessive gas or water from above or below the reservoir
- Increase the driller's ability to inject fracturing fluids

Offshore Drilling. Offshore drilling operations are similar to those on land, but much more expensive. The average offshore gas well is drilled to about 10,400 feet (3,200 meters). A major difference between onshore and offshore drilling is the platform upon which the rig is mounted. An offshore exploratory rig must be able to move across water to different drilling sites.

Offshore rigs include drilling barges, which are used mainly in shallow, protected waters. Jack-up rigs, which have legs that can be raised and lowered, can drill in somewhat deeper water, up to 350 feet (100 meters). A semisubmersible is a whole drilling platform that floats on submerged pontoons and is anchored at the drillsite. Semisubmersibles are very stable during high seas and winds and can drill below water as deep as 2,000 feet (600 meters). Drillships float over the site and drill through a hole in the hull. These can handle virtually any depth of water.

Once a commercial gas field has been discovered offshore, it can be developed with either a fixed or a tension-leg platform. Fixed, steel jacket platforms are the most common. Their legs are held in place by piles driven into the sea bottom. In contrast, a tension-leg platform floats above the offshore field, with small-diameter, hollow steel tubes connected to heavy weights on the sea floor (Fig. 3–4).

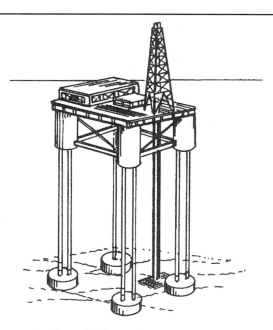

Fig. 3–4 **Tension-leg Platform (Norman Hyne's *Nontechnical Guide to Petroleum Geology, Exploration, Drilling, and Production*, PennWell, 1995)**

Evaluating and Completing a Well

Completing a gas well usually costs more than drilling it in the first place. That's why the well must be evaluated accurately after it has been drilled. Will the well produce enough gas to make it worthwhile to complete? The answer is based mainly on well logs – measurements that characterize the rocks and fluids through which the hole has been drilled.

Well logging

To obtain a physical description of the rocks, geologists perform a lithographic log by sampling the well cuttings (rock chips flushed upward by the drilling mud). A more accurate source of information on reservoir characteristics is a core of the formation extracted from the bottom of the well while drilling is suspended. Core samples can also be taken through the side of the well, which is faster and cheaper.

Other traditional logs include drilling-time measurements, which record the bit's rate of penetration, and mud logs, which analyze the chemistry of the mud and cuttings to detect traces of gas. Mud logging has been improved by development of a sample gas trap that allows more accurate, consistent measurements of gas in mud.

Wireline logs

In the mid-1920s, the Schlumberger brothers (pronounced slumber-jay) and their son-in-law, Doll, developed wireline well logs. In this logging technique, a recording device is lowered into the borehole on a cable, or wireline. The first wireline log measured electrical resistivity (the ability of rocks and fluids to conduct an electrical current). This electrical data was hand-recorded point by point on a strip of paper. Wireline logs helped rotary drilling become popular because they could be run inside a wellbore that had been plastered with clay particles by the rotary drilling process.

In wireline logging, the recording device, called a sonde, is a torpedo-like cylinder filled with instruments that sense the electrical, radioactive, or sonic properties of the formation rocks and their fluids. As the tool is gradually pulled out of the hole, it continuously records these properties along the way. The signals from the device are transmitted through the wireline to the surface, where they are recorded. The log shows the continuously recorded measurements at all points throughout the depth of the well.

Electrical logs indicate the type of rock in the formation by its characteristic resistivity signature. For example, the resistivity of tight sandstones is high, while that of shales is low. Gamma ray logs measure radioactivity, which also indicates the type of rock in the well. The porosity of the rock can be measured by neutron logs or formation density (gamma-gamma) logs. If natural gas is present, the neutron porosity log will read low, and the formation density log will read high. This divergence is called the gas effect.

A caliper log also indicates the type of rock by measuring the wellbore diameter. Strong rocks such as limestones and some sandstones have wellbores about the same size as the drill bit. The orientation of rock layers in a well is determined by dip logs.

The speed at which sound travels through the rock is measured by sonic or acoustic velocity logs. More porous rock transmits sound at slower speeds. However, sonic logs do not capture porosity created by fractures in the rock, which greatly decrease the amplitude of the sonic waves. A sonic amplitude log is needed to detect the presence of fractures.

Until recently, well logging was performed primarily on bare, uncased holes. The reservoir rock's formation density, resistivity, and pressure were difficult to measure accurately once the casing was in place. However, the gas industry is developing logging tools that can perform these same measurements through casing. These logs will help producers develop gas reserves in proven reservoirs by performing tests in existing cased wells.

Other tests

In the 1980s, sensors were developed to generate real-time logs while the well is being drilled, instead of after drilling. These sensors, attached just above the bit on the drillstring, can detect the various types of conventional wireline log data and transmit it to the surface.

A drillstem test is similar to a temporary completion of a well. Gas is allowed to flow up the drillstem to the surface, where it is measured to evaluate the well's performance in that zone.

Well completion

If logging indicates a dry hole with no gas, the well is plugged and abandoned. But if the well looks promising, the bare hole is cased, or lined with metal pipe to seal it from the rock. Most casing is relatively thin-wall, seamless steel pipe, commonly in sections of 30 feet (10 meters). Casing stabilizes the well and prevents the sides from caving in. It also protects any fresh water supplies nearby and prevents the water from diluting the gas during production. Some casing is inserted as the well is being drilled to keep it from caving in.

The final piece of casing is the production string that runs into the producing zone. When the well has reached its target and the last piece of casing has been inserted, wet cement (slurry) is pumped between the casing and the sides of the well and allowed to harden. If the producing formation is composed of unconsolidated sands that can cave into the well, coarse sand called a gravel pack is used to support the formation (Fig. 3–5). Some gas wells are completed in more than one producing zone, called a multiple completion. Multiple completions are common in coal seam gas wells.

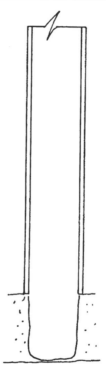

Fig. 3–5 Gravel Pack Completion (Norman Hyne's *Nontechnical Guide to Petroleum Geology, Exploration, Drilling, and Production*, PennWell, 1995)

Once the cement has set, holes called perforations are shot through the casing and cement to allow gas to flow into the well. Then small-diameter tubing is run down the hole to conduct the gas to the surface. The gas flows upward naturally, requiring no pumping. The flow rate of the well is tested and, if production is satisfactory, a series of valves and fittings (called a Christmas tree or production tree) is installed on the wellhead to control the flow (Fig. 3-6). In wells that produce both oil and gas, the two fluids are separated at the wellhead.

When several wells have been drilled successfully, a gathering system is built. This system consists of flow lines collecting gas from several wells and transporting it to a central location for processing.

Natural Gas Production

Pressure on the gas or oil in the reservoir rock causes the fluids to flow from the rock pores into the well. Natural gas is produced from most reservoirs by expansion, where the pressure of the expanding gas underground forces it into the well. Gas expansion produces a relatively large amount of the gas contained in the reservoir. In some reservoirs, water drives the production process. Expansion of water near or below the reservoir causes the gas to flow into the well and then fills the empty pores in the rock. Water-drive reservoirs are not as productive as expansion gas drive reservoirs.

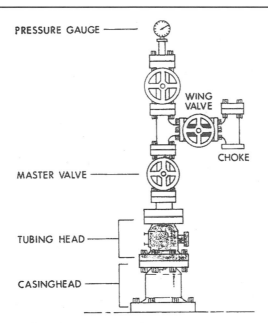

Fig. 3–6 Parts of a Christmas Tree (Norman Hyne's *Nontechnical Guide to Petroleum Geology, Exploration, Drilling, and Production*, PennWell, 1995)

Gas production from a well gradually declines with time as the reservoir pressure decreases. Gas is usually produced from the well until the gas pressure drops to around 700 to 1,000 psi (49 to 70 kg/cm^2), which is the lowest pressure that the pipeline will accept. In some instances, the life of a gas well can be extended by using a compressor to boost the underground gas pressures to pipeline pressure.

At the outset of production, well tests can be run by the well operator, a specialized well tester, or a service company. For example, a productivity test determines the effect of different gas production rates on the reservoir. Also, gas wells are monitored routinely during production to measure the amount of gas, condensate, and water being produced. For fields with a central gas processing plant, production tests also indicate how much each well is producing.

Well stimulation

Many gas-producing formations are "tight," meaning that their permeability is low, which limits the flow of gas through the rock. Several methods can be used to increase permeability, or "stimulate" the well. These treatments enlarge the flow passages in the rock around the wellbore so that the reservoir fluids can flow with less resistance into the well. For limestone formations, acid can be pumped down the hole to dissolve the rock. Wells can also be given a shot of explosives to fracture the rock. Until the late 1940s, fracturing was usually done with liquid nitroglycerin.

In more recent fracturing jobs, a high-pressure liquid is pumped into the well in large volumes to shatter the reservoir rock. This process is called hydraulic fracturing or stimulation. Fracturing fluids include oil, nitrogen foam, water, or water with acid. Once the fractures have been created, they are held open by proppants such as coarse sand. Fracturing jobs are often described by the amount of fluid and proppants used.

Because tight formations are common in gas reservoirs, much of the gas industry's research has focused on stimulation technology. Hydraulic fracturing is very expensive, costing up to $500,000 per well. Computer software has been developed to collect data during a hydraulic fracturing procedure (in "real time") and use the information to predict the growth of the fracture. This allows producers to optimize the process by making adjustments while the fracture is still developing.

Other recent advances include faster, easier fracture mapping, which measures the geometry and final dimensions of the fracture. Wireless tools have also been invented to measure actual downhole temperatures and pressures during fracturing, instead of extrapolating the data from surface measurements.

Natural Gas Treatment

Annually, the gas industry spends about $5 billion to treat natural gas before it enters transmission pipelines. As discussed in Chapter 1, natural gas consists mainly of methane, along with ethane, propane, and butane. In addition, natural gas often contains undesirable impurities such as water, carbon dioxide ("acid" gas) and hydrogen sulfide ("sour" gas). "Sweet" natural gas has no detectable amounts of hydrogen sulfide. Both carbon dioxide and hydrogen sulfide combine with oxygen to form acids that corrode gas piping.

Some natural gas also contains heavier, liquid hydrocarbons that condense when the gas is produced. This "wet" gas can occur as a gas in the reservoir and during production, but it produces liquid condensate on the surface. "Dry" gas is pure methane and forms no liquid either in the reservoir or aboveground.

Gas that has been treated to meet the specifications of a pipeline purchase contract is called pipeline-quality gas. The heating value of pipeline-quality gas is usually around 900 to 1,050 Btu/cu ft (34 to 39 megajoules/m^3). Its pressure must match the typical pipeline pressure of 700 to 1,000 psi (3,400 to 6,900 kPa). Pipeline-quality gas must be dry enough to prevent formation of liquid condensate or hydrates (ice-like particles that can plug pipelines), and it must not contain corrosive gases or excessive moisture.

Improvements in gas treatment technologies are being driven by exploitation of marginal gas resources, which contain greater amounts of impurities, and by more stringent air quality regulations, which require controlling emissions from some treatment processes.

Gas conditioning

Water and impurities in natural gas are removed by gas "conditioning" in the field. Conditioning processes include dehydration to extract water and "sweetening" to remove carbon dioxide (CO_2) and hydrogen sulfide (H_2S). Glycol, a liquid desiccant, is used to absorb the excess moisture from natural gas. Pipeline specifications usually require a maximum water content of 7 lb/million cu ft (kg/m^3).

Glycol dehydration processes emit hazardous air pollutants, primarily benzene, toluene, ethylbenzene, and xylene (collectively known as BTEX). The gas industry has developed an emission control technology that can eliminate 95% of these BTEX compounds in many cases. Also, computer programs have been developed to calculate ballpark estimates of BTEX emissions from specific gas conditioning plants. Using these estimates, the gas producer can conduct actual measurements at those plants most in need of emission controls.

Corrosive gases (CO_2 and H_2S) are removed in a "sweetening" unit that contains iron sponge or other chemical organic bases called amines. When the sponge or chemicals become saturated with acid gases, they can be regenerated with heat and then used again. Pipeline-quality gas requires an H_2S content no greater than 4 parts per million by volume (ppm vol) and a CO_2 content of 1% to 2% by volume (vol %).

About 13% of natural gas produced in the U.S. requires treatment to remove

H_2S, from which sulfur can then be recovered ("scavenged") by the traditional Claus process. This sulfur recovery method, however, is too complex and expensive for gas plants that produce relatively small amounts of sulfur. Recently developed technologies for sulfur recovery include liquid "redox" (reduction/oxidation) and a process based on crystallization of the sulfur.

Because CO_2 can be injected into depleted oil fields to enhance production, it also is sometimes recovered from natural gas and sold as a by-product. For economic reasons, the use of membranes to remove CO_2 from natural gas is growing.

Gas processing

The condensate present in wet gas is recovered in a gas processing plant as a valuable by-product. When removed along with butane, propane, and ethane, the condensate is called natural gas liquids (NGL). These liquids can be recovered by either cooling or absorption. Cooling the wet gas causes the natural gas liquids to separate from the dry gas. This process includes low-temperature separation followed by expansion in a cryogenic or expander plant.

Natural gas liquids can also be removed in an absorption tower by bubbling the gas through trays containing a light hydrocarbon liquid similar to gasoline and kerosene. The natural gas is then removed by distillation. The methane-rich gas remaining after the liquids have been extracted is called tail gas.

Bibliography

Hyne, Norman J., *Nontechnical Guide to Petroleum Geology, Exploration, Drilling and Production*. Tulsa, OK: PennWell Publishing Company, 1995.

4

GAS TRANSMISSION PIPELINE NETWORK

How Natural Gas Is Transported

The pipeline industry carries natural gas from producers in the field to distribution companies and to some large industrial customers. Natural gas transmission pipelines are made of strong, large diameter pipe that operates at high pressures (500 to 1,000 psi, or 3,400 to 6,900 kPa). As gas is transported over long distances, its pressure is maintained by compressor stations sited at strategic points along the pipeline. Gas engines and turbines are often used to power the pipeline compressors.

Pipeline companies that do business within a single state are classified as intrastate. Those that operate larger networks across state borders are called interstate pipelines and are regulated by a federal agency. As of the 1990s, more than 300,000 miles (480,000 kilometers) of gas pipelines criss-cross the United States, serving nearly 60 million gas customers.

Brief History of the Pipeline Industry

As discussed in Chapter 1, natural gas transportation first became profitable when electrically welded, seamless steel pipe was introduced in the 1920s. The strength of this pipe allowed transmission of gas at higher pressures and thus in greater quantities. This technology reduced the cost of natural gas and made it more competitive with other fuels. By 1931, several long-distance transmission systems had been constructed (Table 4–1).

The modern history of the pipeline industry is largely related to its regulation by the federal government. Historically, pipeline companies owned production wells and also purchased gas from producers. The gas was then sold to local distribution companies and, in some cases, directly to large industrial customers. In the 1980s, the Federal Energy Regulatory Commission (FERC) dramatically redefined the function of the industry by requiring interstate pipelines to provide "open access" – that is, transport gas for other customers, even the pipeline's own competitors.

This requirement allowed a local distribution company, a large industrial customer, or a group of customers to purchase gas directly from a producer or marketer and arrange separate contracts for pipeline transportation. The pipeline company's gas marketing, sales, and storage operations had to be divorced from its gas transmission services, which became available to all comers. Effectively, the transportation of gas was deregulated. (The regulatory history of the natural gas industry is discussed in more detail in Chapter 8.)

Since then, competition has driven down the cost of transporting gas nationwide and has forced pipeline companies to focus on reducing expenses. During the 1980s and 1990s, the industry became much more consolidated. Other major concerns of the pipeline industry are public safety and protection of the environment. Interstate pipelines are extensively regulated by other government agencies besides the FERC, which require them to control emissions from compressor stations and other equipment.

Pipeline Project Development

Preliminaries

Before developing a new pipeline, companies must obtain a certificate of public convenience and necessity from the FERC. This usually requires submitting

Table 4–1. **Early Long-distance Transmission Systems**

Company	Gas Field	Original Terminus
Interstate Natural Gas Co.	Monroe, LA	Baton Rouge (1926) & New Orleans (1928)
Canadian River Gas Co. Colorado Interstate Co.	Amarillo, TX	Denver (1928)
Cities Service Gas Co.	Amarillo, TX	Kansas City (1928)
Consolidated Gas Utilities Co.	Amarillo, TX	Enid, OK (1928)
El Paso Natural Gas Co.	Lea Co., NM	El Paso (1928)
Memphis Natural Gas Co.	Monroe, LA	Memphis (1929)
Mississippi River Fuel Corp.	Monroe, LA	St. Louis (1929)
Pacific Gas & Electric Co.	Kettleman Hills, CA	San Francisco & Oakland (1929)
Southern Natural Gas Corp.	Monroe, LA	Atlanta & Birmingham (1930)
Western Public Service Co.	Rock Springs, WY	Salt Lake City (1930)
Atlantic Seaboard Corp.	Kentucky	Washington, D.C. (1931)
Montana Power Gas Co.	Cut Bank, Mont.	Butte, Mont. (1931)
Natural Gas Pipeline Co. of Am.	Amarillo, TX	Chicago (1931)
Northern Natural Gas Co.	Amarillo, TX	Springfield, Ill. (1931) and Detroit (1936)
Southern Fuel Co.	Kettleman Hills, CA	Los Angeles (1931)

construction plans and economic studies that demonstrate a demand for gas in the area to be served and an available, adequate supply of gas. Also, the company must analyze the environmental impact of the pipeline's construction and operation. In 1997, FERC approved three gas pipeline projects to import gas from Canada, two into the upper Midwest and one to northern New England.

Once the federal certificate is granted, the right-of-way must be purchased, and leasing of surface property along the path of proposed construction begins. When a pipeline is designed, economics dictate the choice of gas pressure, pipeline diameter, pipe wall thickness, type of compressors, and compressor station spacing. The goal of all these design specifications is to get the job of transporting a specific quantity of gas per day done safely and at the lowest possible cost. Computer tools have become available to help select pipeline routes and design specifications. Once the design is complete, including all arrangements involving its location, the pipe and fittings are ordered and delivered.

Pipeline construction

The first step in construction is to clear the right-of-way and dig a deep trench using ditching machines. Sections of pipe, or joints, are strung out alongside the trench. Long-distance pipeline diameters usually range from 24 to 48 inches (61 to 122 cm), though some lines are as big as 56 inches (142 cm) in diameter. While tractors and clamps hold the joints of pipe in place, welders form them into a continuous length of line (Fig. 4–1). Long pipeline construction jobs might use automatic welding machines.

Fig. 4–1 Welding Sections of Transmission Pipe

After welding, the outside surface of the pipe is cleaned, coated, and wrapped (Fig. 4–2) to inhibit external corrosion. In some instances, pipe is coated with a film of fusion-bonded epoxy at the mill where it is manufactured, so that only the welded joints need to be coated in the field. Sometimes the inside surface of the pipe is also coated with epoxy film at the mill. Interior

coating has several functions:
- Prevents corrosion during shipment and storage
- Improves reflection of light to facilitate internal inspection of pipe before welding
- Minimizes water retention after hydrostatic testing (see below)
- Reduces absorption of gas odorants during initial operation of the pipeline
- Creates a smooth surface, which improves gas flow efficiency by 4% to 5%

Fig. 4–2 Field Coating and Wrapping

Finally, as the string of pipe sections is formed, it is lowered into the trench (Fig. 4-3), and the trench is re-filled with earth. When rivers must be crossed, the pipe, weighted to prevent flotation, can be lowered into a trench dug in the river bottom, or it might be carried over on a bridge (Fig. 4–4). Sometimes directional drilling techniques (similar to drilling directional gas wells) are used to cross major obstacles. Directional drilling allows the pipeline to be installed underneath rivers, highways, and railroads. It can also be used to avoid damaging environmentally sensitive wetlands, which have become a barrier to pipeline construction in many areas.

Fig. 4-3 Laying Pipeline in Trench

Fig. 4-4 Pipeline Aerial Bridge over the Mississippi River

Pipeline Operation

Transmission lines are costly to build because of the investment in land rights, very large compressors, and huge amounts of high-strength, large-diameter pipe. Because of the size of the investment committed, pipeline companies strive to operate as close to maximum capacity as possible throughout the year. This spreads out the fixed costs of the pipeline investment over a greater volume of gas.

Nevertheless, the demand for gas varies greatly according to the season of the year because large amounts are used for heating during the winter. Also, the market for gas as a power generation fuel is growing, which will change traditional patterns of gas demand.

To minimize transportation costs, the pipeline industry tries to level the rate of gas delivery as much as possible – in effect, to smooth out the winter peaks and summer valleys of gas demand. For example, computer modeling enables the pipeline to look ahead and extract gas from storage just in time when it is needed, rather than having extra unsold capacity built up in the pipeline ("linepack").

Compressor stations

Natural gas is compressed for transmission to minimize the size and cost of the pipe required to transport it. As gas flows through a pipeline, friction inevitably reduces its pressure and flow rate. Thus, the gas must be re-compressed in compressor stations placed at intervals along the pipeline (Fig. 4–5). Usually, stations are sited at intervals of 50 to 100 miles (80 to 160 kilometers). The gas pressure is boosted by compressors rated at several thousand horsepower (kilowatts) each. As of year-end 1995 in the U.S., pipeline compressor power totaled more than 14.0 million horsepower (10 megawatts, MW).

Pipeline compressor stations can use either reciprocating or centrifugal compressors. Reciprocating machines have relatively high compression ratios (the ratio of outlet to inlet pressure) and limited capacities. Typically, they are connected in parallel at a compressor station and are driven by two-stroke, internal combustion natural gas engines. However, some four-stroke gas engines, large electric motors, and in a few cases, steam engines are used to drive reciprocating compressors (Fig. 4–6).

Centrifugal compressors have relatively lower compression ratios and higher capacities. These machines rotate at top speeds of 4,000 to 7,000 rpm and are usually driven by natural gas turbines, though steam turbines and internal combustion engines are sometimes used. Because of their high capacity, compact size, and lower installed cost per horsepower, centrifugal compressors were much in favor when the cost of gas fuel was low.

Fig. 4–5 Typical Pipeline Compressor Station

Fig. 4–6 Reciprocating Engine/Compressor

However, their energy efficiency is lower than that of reciprocating compressors, especially at partial loads. Thus, higher gas costs have tilted the economics of compressor selection in favor of reciprocating compressors whenever the operating hours per year (and thus fuel consumption) are considerable. Currently, reciprocating machines provide more than half of the U.S. pipeline network's total compressor power.

Air pollutants emitted by compressor station engines are one of the pipeline industry's primary environmental concerns. Federal regulations require pipeline operators to reduce emissions of nitrogen oxides from reciprocating engines in transmission service.

Inexpensive technologies have been developed for existing engines to reduce these emissions. The techniques work by mixing the gas fuel and combustion air more thoroughly and by increasing the proportion of air mixed with the fuel. Sometimes, the air and fuel are mixed in special chambers before entering the engine.

The pipeline industry is also developing equipment to control other types of engine emissions, such as toxic or hazardous air pollutants. These emissions are expected to be regulated beginning in 2003. Much of the industry's emission control research is conducted at Colorado State University's engine test facility.

Metering

Metering of gas flow is an important function of pipeline gas operations. Natural gas is measured at the beginning and end of each pipeline section, and often at each compressor station and each spot where the pipeline splits into two lines. Also, some pipelines serve large industrial customers directly, which requires high-volume meters. Orifice meters are the most common type used to measure gas flow in transmission lines, though they are declining in use for meter installations.

The pipeline industry has developed many improved metering technologies to reduce costs and increase the accuracy of gas flow measurement. Many of these technologies are tested at the industry's Metering Research Facility in San Antonio, Texas. Developmental techniques include automated meters, electronic flow monitoring, energy (Btu) measurement, meter calibration, ultrasonic meters, and computerized data acquisition and analysis. A second research facility, in Columbus, Ohio, allows pipeline companies to simulate their own "real-world" operating conditions and evaluate prototype equipment.

Pipeline Maintenance and Safety

Pipelines require regular patrol, inspection, and maintenance, including internal cleaning and checking for signs of gas leaks. The integrity of the pipeline network and its related equipment is one of the industry's top concerns. The threat of a catastrophic pipeline rupture, though extremely unlikely given the industry's safety precautions, hangs over the head of every pipeline executive and employee.

One of the single most important causes of pipeline failures is mechanical damage. This occurs when heavy construction equipment dents the pipe, scrapes off its coating, gouges the metal, or otherwise deforms the pipe in some way. Yet mechanical damage is difficult to prevent. Pipeline companies cannot continuously monitor every foot of the line over thousands of miles to keep people from digging anywhere near it. Many of the pipeline industry's safety programs involve mapping and marking the location of pipe underground to warn people off.

Corrosion is another serious problem plaguing the industry. Corrosion is a sneaky enemy. Until it has caused obvious damage, corrosion is very difficult to detect and locate accurately. Metal pipe corrodes when water or other conditions in the ground create electrical differences between the pipe and the surrounding soil.

Corrosion damage can take many forms, including pitting and cracking. A phenomenon called stress corrosion cracking is especially hard to detect and can be dangerous if left uncorrected. To minimize corrosion, pipeline companies install electrical devices called cathodic protection systems, which inhibit electrochemical reactions between the pipe and surrounding materials.

Another source of pipeline damage is a defect in the pipe's original coating (called a "holiday"). Once the pipe is installed underground, tiny defects can grow into major problem areas.

Pigging

One surefire way to test the integrity of a pipeline is to empty it of gas and pump it full of water at high pressures (a "hydrostatic" test). But obviously, this method is expensive and time-consuming. To avoid unnecessary hydrostatic testing, the pipeline industry has developed various non-destructive methods for inspecting the inside of the pipe. The most important inspection and maintenance tool is called a pig. This is a device that travels through the pipeline, making the characteristic squealing sound that gave it its name.

The earliest pigs were basically pistons that were blown through the pipe to remove dirt and corrosion products, which were expelled at selected points along the line. If not removed, this material increases friction and reduces gas flow rates. Debris also causes wear by erosion of equipment at the end of the pipeline ("downstream"), such as regulators and meters. This type of pig is still in use.

More recently, "smart" pigs have been designed to incorporate technologies for detecting internal pipe conditions that could lead to failures. Many of these smart pigs are equipped with on-board computers for more accurate diagnosis of problem areas. Some pigs contain instruments that measure the wall thickness of the pipe. This indicates spots where corrosion might have eaten away some of the metal. Other pigs can also detect structural abnormalities in the pipe.

However, many older pipelines contain gas valves that are smaller in diameter than the pipe, which prevents smart, computerized pigs from passing through. Some lines also have sharp bends that the pig cannot negotiate. The pipeline industry is working to develop a smart "collapsible" pig that can squeeze itself through smaller diameter valves and turn around corners inside the pipe.

Leak detection

Gas leakage from pipelines can be detected by equipment similar to that used by distribution companies (see Chapter 4). But because of the longer distance of transmission pipelines, many leak surveys are conducted from an airplane. Aerial patrols can see large areas of yellow vegetation, which indicate the drying effect of a gas leak. They can also find areas where the ground has washed out, potentially exposing the buried pipeline, and can observe any construction activities that might threaten to damage the line.

Pipe repair

When leaks or damaged sections of pipe are identified and pinpointed, the traditional method of fixing them is to uncover the pipe, cut out the defective section, and replace it with a new piece of pipe. This method requires shutting down the pipeline temporarily. Alternatively, a metal sleeve can be applied around the pipe section and welded in place. Although the pipeline can continue to operate during this procedure, welding can introduce problems such as cracks and stresses in the welded area.

More recently, the gas industry has started using a less expensive variation of the sleeve repair method. Instead of a metal sleeve, this technology uses a composite wrapping material made of glass fibers impregnated by a resin matrix. This technique is faster and cheaper than welding a metal sleeve, yet provides an equally safe, strong repair.

Bibliography

"Transmission Program Overview," Gas Research Institute, GRI/Net (http://www.gri.org), September 1998.

Albrecht, Jim, "Making 'Smart' Pigs Even Smarter," Gas Research Institute Digest 20, Spring 1997 (No. 1), pp. 12-15.

5

NATURAL GAS STORAGE

How Natural Gas Is Stored

Transmission and distribution companies use many strategies to maintain a steady flow of pipeline gas and match supply with demand. One important method is the development of underground storage facilities, located as near as possible to gas markets. Natural gas can be stored in depleted gas or oil reservoirs. It can also be contained in water-bearing formations called aquifers (Fig. 5–1) and in underground caverns such as salt domes

Gas storage permits pipelines to operate at or near their design capacity, despite seasonal or daily fluctuations in demand. Traditionally, gas has been stored during the summer when customer demand is relatively lower, then withdrawn in the winter when the weather is cold and demand is high. In 1996, about 440 underground storage pools were operating in the U.S. As of late 1998, this storage capacity was 97% full, and the total natural gas inventory amounted to 3.093 Tcf (90 billion m³), a four-year high. The storage facilities are capable of delivering up to 70 Bcf (2 billion m³) of gas per day for short periods.

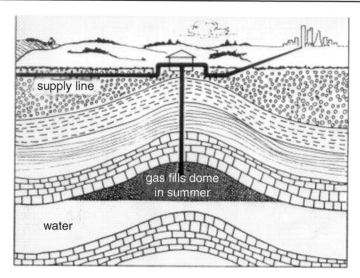

Fig. 5–1 Typical Aquifer Gas Storage Field

Brief History of Underground Gas Storage

The first storage field in the U.S. began operation near Buffalo, New York, in 1916. The Zoar field, a depleted gas reservoir, is still in service. As the gas pipeline network grew after World War II, so did the need for new storage fields to meet gas demand during cold weather. By 1965, the gas industry had developed aquifer storage in the Midwest, as well as deeper depleted gas reservoirs in Pennsylvania, Ohio, and West Virginia.

The first storage cavern in salt formations was created in Michigan in 1961, and the first salt dome storage cavern was opened in Mississippi in 1970. This cavern provided gas to replace production from the Gulf of Mexico that was interrupted by hurricanes.

During the 1960s and 1970s, most storage projects were cancelled due to federal regulation of interstate gas prices. These laws effectively restricted the use of gas to fill storage reservoirs during the summer. In the late 1970s, new federal regulations changed gas pricing again, and more drilling activity led to an oversupply of gas by the early 1980s. Finally, restrictions on the use and storage of gas were lifted.

Options for Gas Storage

Most types of natural gas storage facilities are located underground. Depleted gas reservoirs are the most prevalent storage fields. Gas is "injected" into and withdrawn from these reservoirs by old gas wells. The pressure inside the reservoir rises during gas injection and falls during withdrawal, or production.

A certain amount of "cushion" gas is needed before gas from storage can be produced. Traditionally, natural gas has served as the cushion gas, although the industry is looking at nitrogen and other gases as alternatives. The cushion gas provides enough pressure for gas to flow from the storage reservoir into the well. Cushion gas is not withdrawn during normal operation of the storage reservoir. In contrast, "working" gas can be produced and delivered when needed. This is the volume of gas that is injected and withdrawn during a normal storage cycle.

Aquifers are porous, permeable rock formations underground that are saturated with water. Aquifers offer a gas storage alternative when depleted reservoirs are not available. Several aquifer storage fields are in use in the upper Midwestern U.S. However, aquifer reservoirs require a very high percentage of cushion gas (up to 80% of total gas volume), which limits their usefulness. Also, they have no gas wells or production equipment in place, as depleted reservoirs do, and aquifers must be tested before development, which increases costs.

Caverns are underground cavities that have been mined for coal or minerals or "leached" (carved out by pumping water underground). These include salt caverns (either in bedded salt formations or salt domes) or rock caverns (for example, coal mines). Caverns are ideal for repeated cycling of gas because they act as a pressurized container, and their cushion gas requirements are low (about one-third of the total gas volume). However, leached caverns are expensive to develop because the leaching process involves the circulation and disposal of enormous volumes of water.

Typically, the deliverability of storage reservoirs (their ability to produce gas) tends to decline with time. As gas wells age, they experience downhole damage. Recent gas industry research has developed ways to diagnose declining gas flows and tailor well treatments that are efficient and effective.

In addition to underground gas storage, distribution companies operate aboveground facilities, typically tanks containing liquefied natural gas (LNG) or liquefied petroleum gas (LPG). These supplemental gas tanks usually have much smaller storage capacities than underground facilities (0.5 to 2.0 Bcf, or 10 to 60 million m^3). Aboveground storage tanks provide enough supplemental gas for a limited period, about 5 to 15 days of peak demand. The operation of LNG and LPG storage tanks is discussed in Chapter 6, *Gas Distribution Systems*.

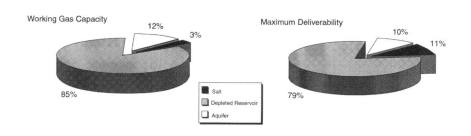

Fig. 5–2 Types of Underground Storage in Use in the U.S.

Gas Storage Customers

Underground fields serve two basic types of service requirements: base load and peak load. Base load storage can hold enough gas to provide the bulk of customer demand beyond the capacity of long-haul pipeline deliveries. Peak load facilities deliver gas at high rates to meet urgent needs for a few hours or days. Peak storage can sustain deliverability for only a limited time, but it can also be replenished quickly, regardless of the time of year. Some depleted reservoirs can be adapted to this type of storage, but the recent trend has been to develop caverns as peak load facilities.

Underground gas storage has traditionally been developed and used by pipelines, but more recently distribution companies have assumed control of more storage facilities. In general, they use storage to secure their supply of gas and meet peak demand. Some gas marketers have also invested in storage fields to convince their customers that they can satisfy contracted commitments.

During the 1990s, demand for new gas storage increased along with gas demand, and storage services expanded. Instead of being a seasonal source of supply that was tapped once a year, gas from storage began to be used more often to boost short-term gas flow rates. Also, changes in federal regulations had ended most of the pipelines' traditional merchant services. Distribution companies and large gas customers became responsible for their own gas supply arrangements, many of which were not long-term contracts.

Innovative types of storage have evolved in response to the changing ways that customers want to use stored gas, from longer term deliveries (150 days) to shorter duration, more flexible services. Most of the older storage fields cannot adjust to faster delivery cycles. As a result, new storage projects have surged during the past five years. More than 80 are scheduled for completion by 2000. Of these, 40 projects could significantly improve the gas system's capacity to deliver higher gas flow rates. The amount of gas in storage could increase by nearly 400 Bcf (11 billion m³), and deliverability could rise by 15 Bcf/day (425 million m³/day).

As peak day natural gas demand continues to grow, so will the need for more flexible gas storage increase. Peak requirements are rising primarily because of demand for residential space heating. Peaking facilities that can store gas for 5 to 20 days will be needed, instead of seasonal storage. Recently, the gas industry has developed techniques to enable old-fashioned reservoir storage fields to provide shorter term service at higher gas flow rates. These re-engineering techniques focus on improved diagnostic software for reservoir characterization, as well as analysis of reservoir management issues.

Bibliography

Albrecht, Jim, "Underground Gas Storage: Improving the Process, Enhancing the Resource," *Gas Research Institute Digest* 21, Summer 1998 (No. 2), pp. 12-15.

Beckman, Kenneth L., and Determeyer, Peggy L., "Natural Gas Storage: Historical Development and Expected Evolution," *GasTIPS*, Spring 1997 (No. 2), pp. 13-22.

Ewing, Terzah, "Strong Dose of Winter Spikes Price of Natural Gas, Reversing Recent Losses," *Wall Street Journal*, November 6, 1998, page C17.

CHAPTER

6

GAS DISTRIBUTION SYSTEMS

How Natural Gas Is Distributed

A distribution system is a piping network that carries natural gas to its customers from various sources of gas supply (Fig. 6–1). The piping network consists largely of service lines that enter the customer's home or business and gas distribution mains, which are larger diameter pipes that transport the gas underground throughout the area served by the distribution company. Many of the older distribution systems in the U.S. were constructed with metal pipe, but most mains and service lines being installed more recently are made of plastic pipe.

The gas distribution industry spends more than $60 billion a year, most of it on purchasing gas supplies. Cross-country pipelines and gas storage facilities are the primary sources of natural gas for distribution. During the winter and at other times when demand for gas is high, the distribution system can also be supplied with liquefied natural gas and liquefied petroleum gas. These supplemental sources of gas cost more than everyday natural gas supplies.

Local distribution companies are often referred to as LDCs or utilities. The terms are virtually synonymous. Municipal utilities function as a gas distribution company for a city or metropolitan area.

Brief History of the Distribution Industry

The first gas company in the United States was founded by four businessmen in Baltimore in 1816. By the 1990s, more than 1,200 distribution companies were providing gas service to nearly 60 million customers in all 50 U.S. states. In the early years, construction, operation, and maintenance of pipes and other facilities were performed by the "street department". Although the materials, equipment, and technology used for gas distribution have changed dramatically, the distribution industry continues to have the same basic responsibilities.

Customer service consisted of installing piping on the customer's premises. Appliance repair was a very important service, because no one else had these skills. Customer service was also responsible for installing, reading, and repairing meters. These functions still exist in gas distribution companies, but the extent of

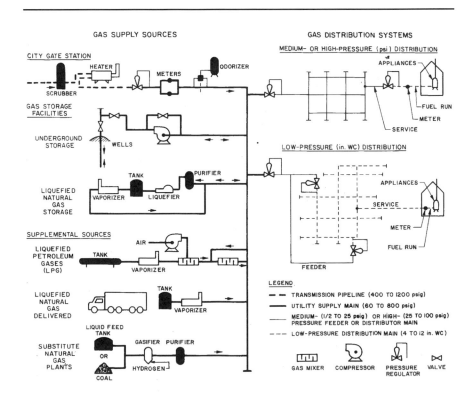

Fig. 6–1 Natural Gas Distribution System

the responsibilities varies widely from company to company.

The last half of the 20th century has been a period of dynamic change for the gas industry. The three segments of the industry – production, transmission, and distribution – can still be discerned, but their responsibilities and the relationships among them and other players have changed. Through mergers and acquisitions, many distribution companies have grown into vertically integrated corporations, while others have diversified to provide other types of utility service and non-utility functions. These changes are described in more detail in Chapter 8, *Regulatory History of Natural Gas*.

Receiving Gas Supplies

City gate stations

Natural gas is received at "city gate" stations, also called town border or tap stations (Fig. 6–2). After gas arrives at the city gate, it is sometimes passed through a cleaner to remove liquids and dust. A primary function of the city gate

Fig. 6–2 City Gate Station

station is to measure (meter) the volume of incoming gas. Most city gate stations measure the gas with orifice meters, though other types of meters might be used alone or in combination with orifice meters.

Natural gas is delivered to the city gate at high pressures, which are required for pipeline transmission, but the distribution system requires much lower pressures. Thus the other main function of the city gate station is to reduce the pressure of the gas. Mechanical devices called pressure regulators lower the gas pressure and control its flow rate to maintain desired pressure levels throughout the distribution system. As the pressure is reduced, the gas also becomes much cooler. For this reason, the gas might be heated to prevent the formation of frost, ice, or ice-like hydrates in the piping and frost heaving in the ground surrounding the pipe.

Deliveries of gas to the distribution system are monitored by a gas dispatching system. Many companies use supervisory control and data acquisition (SCADA) systems, which provide on-line data and can rapidly respond to changing conditions in the distribution system.

Natural gas has very little odor, especially after processing in the field, so odorization is a very important step in the distribution process and is required by federal safety regulations. If the pipeline gas is received with insufficient odor, it must be odorized before leaving the city gate station. The odorant confers a "gassy" smell that makes the presence of escaping, unburned gas recognizable at very low concentrations. This warns customers well before the gas can accumulate to hazardous levels. Mixtures of air and natural gas are explosive over the range of 5% to 15% natural gas. To ensure safety, odorized gas is detectable at a concentration of just 1%.

Supplemental gases

Although most natural gas is delivered via pipeline, distribution companies keep supplemental gas supplies on hand. Supplementary gas is generally produced only at times of highest demand (peak load), such as on cold winter days. In contrast, the normal level of gas demand is called the base load. Producing and distributing supplemental gas is called peakshaving. Even though gas for peakshaving is more expensive than pipeline gas, it is still less costly than buying additional firm supplies of pipeline gas in advance of the heating season.

The two dominant sources of peakshaving gas are liquefied natural gas and liquefied petroleum gas (LNG and LPG). Because these fuels are in liquid form, they are more compact and easier to store than gaseous natural gas. LPG consists mainly of propane and butane. In a distribution company's LPG peakshaving plant, propane or butane is warmed and vaporized, mixed with air, and then injected into the distribution system's gas stream. Mixing with air makes the LPG compatible with the customer's natural gas appliances.

In an LNG plant, the liquid gas is stored in tanks at very low (cryogenic) temperatures to keep the fuel in its liquid state. When needed to meet peak demand, the LNG is warmed and vaporized. Some LNG plants have their own gas liquefaction equipment. Natural gas from the pipeline is liquefied during periods of low demand on the system. Other LNG plants have only storage tanks and vaporization equipment. The LNG is purchased from a supplier and transported to the plant in cryogenic tanker trucks. Not all LNG is domestically produced in the U.S. LNG from foreign sources is shipped in specially designed vessels to terminals along the American coast, where it is stored for delivery.

During the 1970s, the U.S. experienced a period of natural gas shortages. Many companies installed plants to produce substitute natural gas (SNG) from naphtha, a by-product of oil refining. This SNG was used to provide base load natural gas to large customers who had been cut off, or curtailed. However, SNG is expensive to produce, and these plants are no longer used for base load gas supply.

Distribution System Operation

Piping and pressure regulation

The pipe that receives gas at the city gate station and carries it into the distribution system is called a supply main. In some cases, this might be just a few hundred feet in length. In other instances, the supply main can consist of many miles of complex piping. The pressure of gas in the supply main is lower than the transmission pipeline pressure, but higher than that of the distribution system. Supply mains might have a few high-pressure service lines connected directly to large industrial customers.

In addition to supply mains, four other types of piping are used in distribution systems:

1. Feeder mains transport gas from the pressure regulator or supply main to the distribution mains. Feeder mains might also have some service lines connected to large industrial customers.
2. Distribution mains supply gas primarily to residential, commercial, and smaller industrial customers.
3. Service lines deliver gas from the distribution main in the street to the customer's meter. Service lines are usually the property and responsibility of the utility. However, some utilities own only the portion of the service lines in the public right-of-way, and customers own the portion on their property.

4. Fuel lines are customer piping beyond the meter to appliances. These lines are the property and responsibility of the building owner.

Many gas distribution systems consist of several superimposed networks of mains operated at different pressure levels. The actual pressure in the mains fluctuates from day to day and hour to hour. Some utilities deliberately set the pressure at higher levels in the winter, when demand is greater. Figure 6–3 shows the typical operating pressures of different segments of the system.

Large industrial customers served directly from the supply main require specially designed meter and regulator sets to handle large gas volumes and high delivery pressures. Smaller commercial and residential customers who are served directly from a high-pressure supply main require an additional regulator to reduce the pressure from the service line. These services are sometimes referred to as farm taps, a term coined from the gas transmission business.

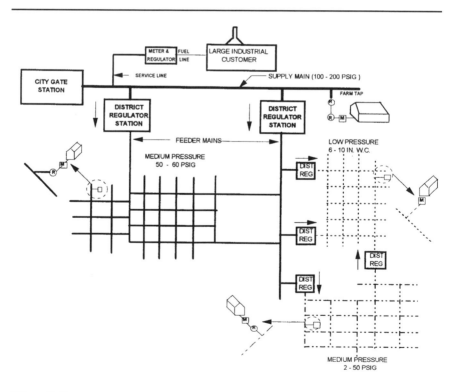

Fig. 6–3 Operating Pressures in a Distribution System

Feeder mains deliver gas into the distribution networks. Newer systems typically operate at a pressure of 60 psig (415 kPa), which requires a service regulator on each service line to reduce the pressure to levels acceptable for consumer appliances. If the distribution network operates at lower pressures, district regulator stations reduce the pressure of the gas from the feeder main before it enters the system.

Many older distribution companies, usually in big cities, have large segments of piping that operate at very low pressures (also called standard or utilization pressure), in the range of 6 to 10 inches water column (0.2 to 0.3 psig or 1.5 to 2.5 kPa). This piping originally distributed gas from manufactured gas plants. Household appliances are generally designed to operate near this pressure level, at about 4 inches water column (0.1 psig or 1 kPa), so individual service regulators are not required. However, pressure regulators are often built into appliances such as ranges, water heaters, and furnaces to ensure good performance and allow fine adjustments to the incoming gas pressure.

Classification of piping networks by pressure level is quite common, but utilities do not agree on the terminology or pressure range covered by each classification. Federal safety standards define a high pressure distribution system as one in which the pressure is higher than that supplied to the customer (that is, a service regulator is required to reduce the pressure), while a low pressure system is one that operates at substantially the same pressure as supplied to the customer.

Controlling gas leaks

Distribution companies perform leak surveys to ensure that all gas piping (including that on customer property) is inspected regularly and shown to be leak-free. Gas can be detected both above and below ground by extremely sensitive leak detectors. Gas company employees on leak patrol carry portable units or drive a leak detection vehicle to survey longer sections of pipe.

Also, escaping gas has a drying effect on vegetation, so the effects of leakage are visible where the vegetation changes color from green to brown or yellow. For inspecting buried pipe, distribution companies dig small holes at locations indicated as suspicious by an aboveground survey or a reported gas odor. Conditions inside the piping are monitored with special cameras.

Distribution service areas are divided up and surveyed according to a schedule, with special attention given to public buildings such as schools, hospitals, and theaters. Leaks are classified according to their intensity and location, and those with the highest priority are repaired immediately. Distribution companies are also responsible for promptly investigating any reports of an odor, leak, explosion, or fire that might involve a gas line.

Distribution System Construction

Gas distribution mains are installed in undeveloped land, such as new housing projects, by digging a trench and burying the pipe. Sometimes connections for other services, such as phone lines and cable television, are installed along with the gas pipe. For new mains and services, polyethylene plastic pipe is used almost exclusively. In contrast, older distribution systems were constructed of cast iron or steel piping. The diameter of gas main piping usually ranges from 1.25 to 3 inches (3 to 7 cm) but can be as large as 6 inches (15 cm). Service lines are typically 0.5-0.75 inches (13 to 19 mm) in diameter.

Plastic pipe is often delivered to the construction site in huge coils up to 500 feet (152 meters) long. Large diameter plastic pipe comes in 30 to 40-foot (9 to 12-meter) sticks. Most of the time, sections of pipe are connected by heat fusion, where the two ends are melted and forced together. Tracer wire for locating the buried pipe later is installed along with the pipe. Before the trench is filled, the pipe is padded with sand or other clean fill.

In developed areas, pipe installation is much more difficult due to pavement such as streets and sidewalks, as well as people's lawns and driveways. Trenches are still dug for most mains, but "trenchless" technologies such as directional boring (drilling) are growing more popular. A horizontal hole is bored underground, and the pipe is pulled through the hole instead of being laid in a trench. Horizontal boring minimizes disruption of traffic and local business, improves public relations by causing less inconvenience to customers, and reduces the utility's cost of restoring the surface to its original condition. Trenchless methods are also used to rehabilitate old mains. For example, plastic piping or liners can be installed inside an aging cast iron pipe.

Distribution System Maintenance

One of the distribution companies' biggest jobs is maintenance, primarily repairing mains and services. The first step in repair is locating the pipe. Pipe location also includes marking gas distribution piping for construction companies, cable television installers, and other third parties. Distribution companies are fully responsible for the cost of repairing any third-party damage that results from inaccurate pipe location.

Compared to metal piping, plastic pipe is cheap and durable, but it is also harder to find underground. Typically, a radio signal is induced in the tracer wires buried alongside the pipe. A hand-held receiver is moved along the spot where the pipe is supposed to be, until the signal reaches its maximum intensity. Tracer wire systems have proved to be quite reliable in locating buried pipe to within 18 inches. In addition to tracer wires, some distribution companies use maps of the system to locate pipe, and a few companies install electrical pipe locating systems.

Leak repair is another critical maintenance responsibility of the distribution company. Corrosion of metal pipe is the number one cause of underground leaks in gas mains. However, leaks caused by third-party damage are also important because the pipe is usually ruptured and the safety hazard is greater.

In most cases, plastic mains that leak are repaired rather than replaced, whereas leaking metal pipe is often replaced with plastic. To repair a leak in plastic pipe, the gas flow is shut off and the pipe is squeezed off on either side of the leak. Then the damaged section is cut out and a new section installed. Cast iron or steel pipe can also be repaired, but the process is more difficult and expensive.

Once the leak has been repaired, the distribution company must restore the site to its original condition. After the soil has been put back into the hole, it must be compacted thoroughly to make sure that it does not settle and sink after restoration is complete.

Other Distribution Activities

Besides delivering gas to customers, distribution companies perform a variety of services, and many utilities have even entered other businesses such as selling gas appliances. Essential customer services include meter installation, meter reading, and turning gas supply on and off. Distribution companies spend more on operating customer meters than they do on operating mains and performing surveys. Record keeping of customer accounts is another function performed by distribution companies. These billing activities actually cost about twice as much as maintaining the distribution piping network. However, computerized systems are rapidly being adopted to automate as many of these functions as possible.

Bibliography

npb associates, *Distribution Survey: Costs of Installation, Maintenance and Repair, and Operations*. Gas Research Institute Report, www.gri.org.

7

USES FOR
NATURAL GAS

Natural Gas Consumption

A ll of the gas industry operations discussed so far have a single purpose – the safe delivery of energy in the form of natural gas for use in appliances and other gas-burning equipment. Gas customers, or end users, fall into three broad classes: residential, commercial, and industrial. In addition, utilities and other power suppliers use gas to generate electricity, and some fleets of vehicles use gas as a transportation fuel.

Homeowners and other residential customers use gas for heat, hot water, cooking, and clothes drying, as well as in gas fireplaces and logs. Although residential customers in the U.S. account for the greatest number of gas users, they consume less than one-fourth of total gas volumes delivered. Commercial businesses, which account for about 14% of gas consumption, use gas mainly for heat and hot water. Some large buildings are also air-conditioned by gas-powered equipment, and many restaurants cook with gas. Industrial customers, which represent about 44% of all gas consumed, use the fuel in tens of thousands of factories and mills to manufacture a great variety of products, from paper to cars.

In the early 1970s, the energy crisis forced industrial countries to reduce their consumption of oil, gas, and other energy resources. To conserve heating and cooling energy, buildings were constructed with less ventilation. Nearly all industrial processes were modified to become more efficient. Also, greater concern for the environment emerged in the U.S., resulting in laws to promote clean air and water. Although energy prices have moderated, current gas usage patterns, and the appliances and other equipment in which gas is consumed, reflect this profound shift toward conservation. Trends in gas consumption are discussed in Chapter 10, *Future Supply and Demand for Natural Gas*.

Inside buildings, piping carries natural gas to each appliance or piece of equipment. Traditionally, rigid pipe made of cast iron ("black" iron) or steel has been used in these interior distribution systems and is still the most common type. However, gas companies have recently begun encouraging builders to use flexible tubing made of copper or corrugated stainless steel to distribute gas inside buildings (Fig. 7–1).

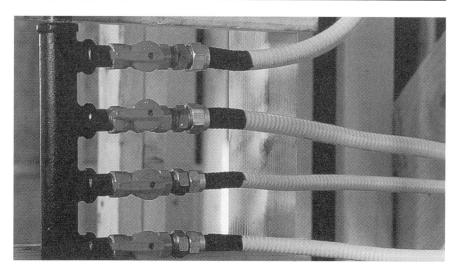

Fig. 7–1 Flexible Gas Tubing (Photo courtesy of Gas Research Institute)

Although these materials are more expensive, they are faster and easier to install, so that savings in labor costs can make up for the difference in price. Also, some homes are equipped with gas "outlets" that are similar in function to an electrical outlet for plugging in appliances. These outlets make it more convenient for people to relocate gas appliances or remove them for cleaning or repair.

Many rural areas are not served by a gas distribution company or pipeline. In these areas, homeowners often maintain their own tanks of liquid propane or lease them from fuel suppliers. Nearly all gas appliances can be adapted to use propane instead of natural gas.

Residential Uses

Heating is by far the most important use of gas in America's residences, with water heating being the second largest. Single-family homes typically use a central heating system powered by a gas warm-air furnace (Fig. 7–2). In fact, these furnaces account for more than 80% of all residential gas heating equipment sales. In 1992, federal regulations required manufacturers to produce furnaces with a minimum fuel efficiency of 78%. But even before then, the efficiency of gas furnaces was being increased, and many models entering the market by the 1990s were well over 90% efficient. Most products use a fan to draw air into the furnace; "induced" draft fans pull the air through, while "forced" draft fans push it. Recently, furnaces have been introduced with "modulating" burners, which maintain a steady temperature by adjusting their heat output gradually instead of cycling on and off (Fig. 7–3).

Natural gas is also used in residential boilers, especially in the Northeast,

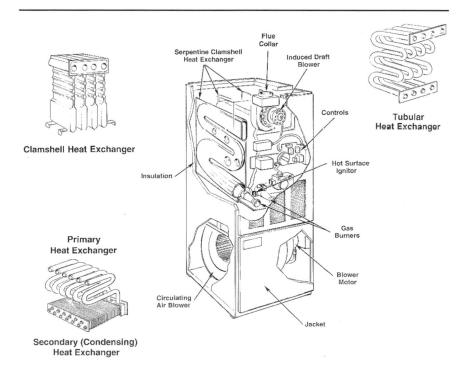

Fig. 7–2 Anatomy of a Gas Furnace (Reprinted with permission from *GATC Focus*, a publication of Gas Research Institute)

which produce steam or hot water for circulation through radiators or hydronic piping. In warmer regions of the country, gas space heaters and wall heaters are more common. Gas space heaters can be picked up and moved around almost like an electric heater, and gas wall heating systems (Fig. 7–4) are catching up with furnaces in terms of efficiency.

Fig. 7–3 **Rheem Comfort Control Modulating Gas Furnace (Photo courtesy of Gas Research Institute)**

Like furnaces, gas water heaters were required by federal regulations to reach higher efficiencies ("energy factors") beginning in 1990. Figure 7–5 shows a typical residential gas water heater. Gas water heaters usually cost less to operate than electric models. Outdoor water heaters, which can be installed in the backyard, are available for moderate climates.

Residential gas cooking appliances have been improved with sealed burners, easier cleaning features, and multifunction ovens for greater cooking flexibility. Newer gas ranges and ovens are also much more sleek and fashionable, or European-looking, than just a decade or two ago. Small gas ranges have been developed for use in apartments and kitchens with limited space, and many gas ovens offer a self-cleaning option similar to electric ovens.

Recently, a variety of gas hearth products have also been introduced, including gas logs, "inserts" for retrofitting wood-burning fireplaces, free-standing stoves, and complete, all-gas fireplaces (Fig. 7-6). Several manufacturers also offer "vent-free" gas fireplaces that need no chimney or exhaust piping.

Fig. 7–4 Empire Wall Furnace (Photo courtesy of Gas Research Institute)

During the 1990s, sales of gas hearth products increased dramatically, especially in urban areas with poor air quality, where burning of wood is restricted. Specialty dealers of hearth products pioneered this growth, but some local gas distribution companies and contractors sell hearth products as a sideline business.

During the 1990s, new homes in the higher price range often contained more than one gas fireplace, and hearth products are being developed to supplement the home's heating system. The gas industry and hearth product manufacturers expect all of these appliances to continue selling well. Eventually, gas fireplaces and other hearth products could consume more gas than cooking appliances do.

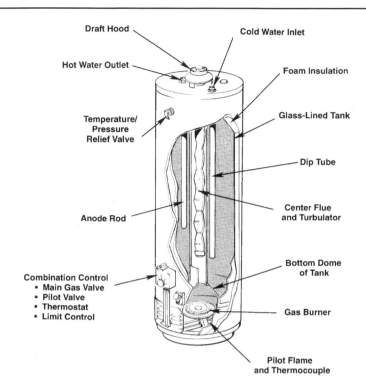

Draft Hood

Cold Water Inlet

Hot Water Outlet

Foam Insulation

Temperature/
Pressure
Relief Valve

Glass-Lined Tank

Dip Tube

Anode Rod

Center Flue
and Turbulator

Combination Control
• Main Gas Valve
• Pilot Valve
• Thermostat
• Limit Control

Bottom Dome
of Tank

Gas Burner

Pilot Flame
and Thermocouple

Fig. 7–5   Anatomy of a Gas Water Heater   (Reprinted with permission from *GATC Focus*, a publication of Gas Research Institute)

Fig. 7–6   Gas Fireplace   (Photo courtesy of Gas Research Institute)

Commercial Uses

Commercial gas customers run the gamut from small to large, including hotels and motels, fast-food outlets and full-service restaurants, convenience stores and groceries, hospitals and nursing homes, schools and universities, shopping malls, discount stores, laundries, office buildings, and warehouses. As in residences, gas in the commercial sector is used primarily for heating.

In small commercial applications, many gas furnaces, boilers, and water heaters are very similar to residential ones, just larger in capacity. In addition, gas heating systems are often installed on the roof of commercial buildings. Rooftop units are also referred to as "unitary" or "packaged," as opposed to large, central heating systems. Like furnaces, rooftop units have become more efficient and easier to install. Some manufacturers offer models with modulating gas burners (Fig. 7–7).

Fig. 7–7 **Trane Rooftop Heating Unit (Photo courtesy of Gas Research Institute)**

Water heating is the second largest commercial gas use. More than 80% of all commercial water heaters sold in the U.S. are powered by natural gas. Gas water heating is especially prevalent in laundries, beauty salons, and other commercial applications that use large volumes of hot water. In restaurants, gas "booster" water heaters are used to raise the dishwashing water temperature to the level required by government regulations.

Gas cooking in restaurants, hospitals, and schools is growing in popularity and accounts for the third largest use of gas in the commercial sector. Many kitchen chefs will not cook on anything but gas, and some commercial gas ranges have been adapted to residential use in expensive homes. Specialty gas cooking products for commercial restaurants include combination ovens, "clamshell" (double-sided) broiler/griddles, and convection ovens. Gas cooking appliances have also been developed especially for fast-food restaurants, including deep-fat fryers (Fig. 7–8) and griddles. The gas industry has made these appliances much easier to control so that fast-food employees need very little training to operate them.

Fig. 7–8 **Frymaster Deep-fat Fryer (Photo courtesy of Gas Research Institute)**

Gas cooling

Unlike the residential sector, many large commercial buildings use gas for air conditioning as well as heating. Central chilled-water systems are common in hospital complexes, university campuses, and office buildings. Most gas-powered chillers use an absorption process to produce cold water for air conditioning (Fig. 7–9). These chillers are recognized for their quiet, troublefree operation. In contrast to electric chillers, gas absorption units use water as the refrigerant instead of chemicals that cause global warming. Because of government restrictions on the use of these chemicals, some electric chillers are being replaced with gas-powered units.

The other method of commercial gas air conditioning is the engine driven chiller. Natural gas engines have been used for years to power the cooling process, but only recently have compact, high-efficiency packaged systems become available and gained popularity. The concept of gas engine driven cooling is simple – the engine just replaces the electric motor in a conventional chiller. The engine supplies power to the compressor, which produces chilled water. Gas engines have some advantages over electric motors in air conditioning applications. For example, the engine can operate over a range of speeds, so its output can match the building's need for cooling more efficiently. Also, the engine gives off heat that can be recovered and used to produce hot water for other building needs.

Fig. 7–9 York Millennium Gas Absorption Chiller (Photo courtesy of Gas Research Institute)

In general, gas-powered chillers cost more to buy than conventional electric machines but cost less to operate. Electric utilities charge commercial customers much higher prices during the daytime and during the summer. Despite widespread publicity about deregulation of the market for electricity, most industry experts to not expect these "peak demand" charges to become substantially lower.

By reducing or eliminating peak demand charges, gas-powered chillers can minimize the commercial customer's energy costs and utility bills. Some commercial building operators are installing "hybrid" air conditioning systems, which use a mix of gas and electric chillers. The choice of which units to operate depends on the gas and electric rates at any given time. Usually, the gas chillers would run during the day when electricity prices are high, and the electric units would run at night during "off peak" periods. Recently, one chiller manufacturer introduced a hybrid unit that combines a gas engine and electric motor in a single package.

In addition to air conditioning, the term "gas cooling" includes refrigeration, which requires lower temperatures. Like chillers, refrigeration systems can be powered by a gas engine. The advantages of engine driven refrigeration are similar to those of gas air conditioning, mainly reduction of electric bills. Custom built systems have been available for years but, as with chillers, have recently been packaged into standard models.

Gas engine driven refrigeration first found a market in ice production, ice rinks, wineries, breweries, meat packing, and cold storage (refrigerated warehouses). During the 1990s, it began gaining popularity in other food processing applications.

Dehumidifying ventilation air

A growing use of natural gas in the commercial sector is for dehumidifying air from the outdoors before it enters the building's heating and cooling system. As recently as the 1980s, commercial buildings were being constructed with very little ventilation. To conserve energy, the indoor air was recirculated throughout the building, mixed with very little fresh air from the outdoors. However, people soon realized that this practice impaired the quality of the indoor air, making it feel stale or stuffy. Also, poor ventilation often led to "sick building syndrome," where chemicals in the carpet and other furnishings lingered in the recirculated air and caused headaches and other symptoms of illness.

Recently, the heating and air conditioning industry changed its official standards for ventilation air. Buildings are required to introduce three times more fresh air for ventilation, in order to improve indoor air quality. However, electric air conditioning systems are not designed to handle these larger volumes of outdoor air, especially in areas with high humidity. Conventional electric air conditioners remove moisture by condensing it, which requires cooling the air to much lower temperatures and then reheating it to comfortable levels. Obviously, this is a very inefficient process.

That's where gas desiccant systems come into the picture. The gas industry had been developing systems for dehumidifying air for many years. These systems use a desiccant (drying) agent such as silica gel to remove moisture from the air by adsorption. Natural gas heat can be used to "regenerate" the desiccant, or restore its ability to adsorb moisture. These early gas dehumidification systems were first used in supermarkets and hospitals, where condensation of water in air conditioning systems caused expensive problems or even dangerous conditions indoors.

With the revised ventilation standards, gas desiccant systems are becoming more common, particularly in humid climates. Typically, they are used to "precondition" ventilation air (adjust its humidity and temperature) before it enters the air conditioning system. Innovative products have been developed and introduced, including simple ventilation modules and more complex systems that recover energy from the building exhaust and use it to regenerate the desiccant

(Fig. 7–10). Some of these ventilation products are designed to bolt onto conventional electric rooftop air conditioners.

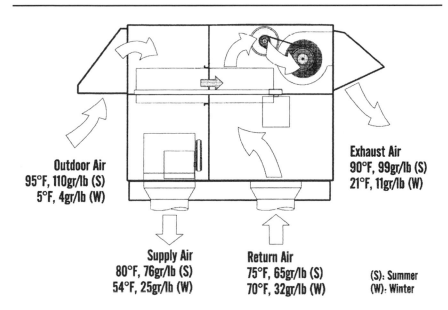

Outdoor Air
95°F, 110gr/lb (S)
5°F, 4gr/lb (W)

Exhaust Air
90°F, 99gr/lb (S)
21°F, 11gr/lb (W)

Supply Air
80°F, 76gr/lb (S)
54°F, 25gr/lb (W)

Return Air
75°F, 65gr/lb (S)
70°F, 32gr/lb (W)

(S): Summer
(W): Winter

Fig. 7–10 Schematic of Desiccant Energy Recovery System (Courtesy of SEMCO Inc.)

Industrial Uses

In the industrial sector, gas is used in a great variety of different processes. Like residential and commercial users, industrial customers use gas for heating and cooling their factories and mills. But huge volumes of gas are also consumed in industrial furnaces, boilers, melters, dryers, and other manufacturing equipment. Industrial applications for natural gas are limited only by the ingenuity of engineers and manufacturers.

The industrial sector of the U.S. economy is classified into dozens of categories, but most gas is used to supply process heat and steam, which are used in metal melting, treating, forging, and casting; curing and forming of plastics and glass; drying of paper, textiles, paint, and coatings; glass melting; and other industrial processes. Natural gas is also used as a "feedstock," or raw material, in producing chemicals from petroleum, such as gasoline additives.

Steelmaking and metal melting

Over the last two decades, the iron and steel industry has begun recycling more scrap metal, which accounts for about 40% of steel production in the U.S. Pieces of scrap metal are melted in electric arc furnaces that are very expensive to operate. To supplement the electric heat, steelmakers use high-temperature natural gas/oxygen burners, which increase the efficiency and productivity of the scrap melting process. These "oxy/gas" burners (Fig. 7–11) also help eliminate cold spots inside the electric furnace. By the 1990s, more than one-fourth of all U.S. electric arc furnaces used oxy/gas burners. They have also become prevalent in many other industrial processes, such as glass melting.

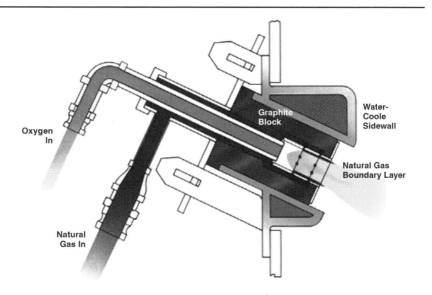

Fig. 7–11 Oxy/gas Burner (Courtesy of Gas Research Institute)

Natural gas is also used as a supplementary fuel in blast furnaces, which melt iron ore to produce pig iron, the main ingredient of steel. Coke (a by-product of coal) is the primary blast furnace fuel, but when a mixture of natural gas and oxygen is injected into the furnace, less coke fuel is needed. Gas injection reduces the cost of producing iron and increases the productivity of blast furnaces.

Gas is also used in steel mills to improve product quality. For example, as hot strips of steel come off of the mill's rollers, the edges of the steel cool down more quickly than the center. Gas is used to equalize the temperature of the strips as they are produced. This process is called continuous strip heating or annealing. Similarly, gas burners are used inside steel reheat furnaces to eliminate cold spots on steel products.

Special natural gas burners are also used to produce aluminum and other non-ferrous (non-iron) metals. To save energy, all aluminum manufacturers in the U.S. use "regenerative" gas burners, where pairs of burners fire in an alternating on/off sequence. The "off" burner recovers heat from the "on" burner's exhaust gas and uses it to preheat combustion air, which reduces fuel consumption (Fig. 7–12). Every 20 seconds or so, the process reverses. Regenerative burners are also used in many other industrial processes, such as casting and heat treating.

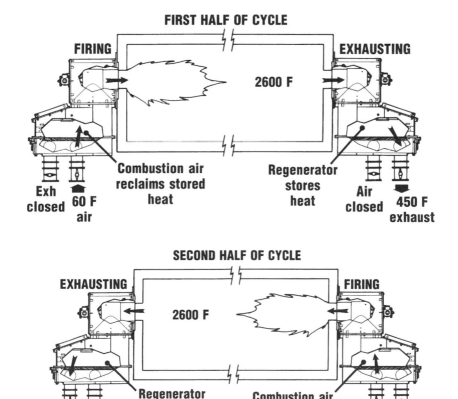

Fig. 7–12 Regenerative Burner (Courtesy of North American Manufacturing Co.)

Heat treating

Nearly everything that's manufactured uses some sort of heat-treated product. More industrial furnaces are used for heat treating than for any other manufacturing process. In fact, heat treating furnaces outnumber movie theaters in America by a factor of about three to one. The purpose of heat treatment (defined as the controlled heating and cooling of a metal or alloy) is to give the material certain desired properties. For example, metal products are heat treated to make them harder and more resistant to wear and tear.

Natural gas has traditionally been the heat treater's favorite fuel. In addition to heating the product, gas is used to create a "controlled atmosphere" inside the furnace, which can either protect the metal from oxidation or promote chemical reactions that improve the product. Most heat treating furnaces are the controlled atmosphere type, and many use regenerative gas burners to improve heating efficiency. Some smaller heat treating factories use radiant gas burners for their furnaces. These burners heat the product much more quickly than electric burners, and they recuperate, or recover, heat for higher efficiency. Radiant gas burners can be made of metallic or ceramic materials and are used in other industrial processes as well as heat treating.

A second type of heat treating furnace creates a vacuum inside, which protects the metal from oxidation; then other elements (carbon or nitrogen, for example) are injected if needed to improve the product. Vacuum heat treating provides a better quality product than controlled atmospheres, but at a higher price. Although vacuum furnaces are much less common than the controlled atmosphere type, they have been growing in popularity. Until recently, nearly all vacuum heat treating furnaces were heated electrically. In the late 1980s, natural gas fired vacuum furnaces were introduced. Since then, gas vacuum technology has been improved to reach the higher temperatures required for many heat treated products (over 1900°F, or 1040°C).

Glass manufacturing

Natural gas is the predominant fuel used by the glass industry. Glass is melted at temperatures up to 2800°F (1540°C) in huge gas-fired furnaces that produce several hundred tons of glass per day. These furnaces contain bricks ("checkers") that capture the furnace's exhaust heat, which is then used to preheat combustion air. Many glass furnace operators have installed oxy/gas burners to further improve process efficiency. However, because of the high process temperatures required, glass melting furnaces produce large amounts of nitrogen oxides. Although these emissions are not yet regulated nationwide, the glass industry is being forced to reduce nitrogen oxides emissions in southern California.

One way to do this is called oxygen-enriched air staging (Fig. 7–13).   This process works by initially starving the burner flame (reducing its oxygen), which inhibits formation of nitrogen oxides.   Then later in the process, oxygen-enriched air is injected to burn off any remaining fuel or carbon monoxide.   Existing glass furnaces can be easily adapted to use the oxygen-enriched air staging process by adding oxygen flow meters and injectors.   The gas industry is also investigating a technique called oscillating combustion to further reduce nitrogen oxides emissions.

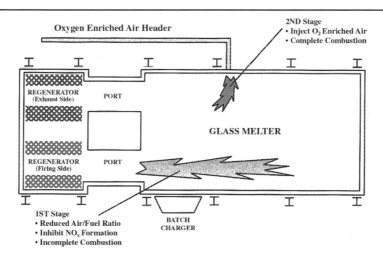

Fig. 7-13 Oxygen-enriched Air Staging in a Glass Melting Furnace (Courtesy of Combustion Tec)

Raw glass materials ("batch") or glass scrap ("cullet") can also be preheated before they are melted.   Preheating reduces the amount of fuel and oxygen required in the overall melting process. The equipment recovers waste heat from the glass melting furnace and uses it to preheat the glass to temperatures of 900° to 1100°F (480° to 590°C).

After melting the glass, natural gas can also be used to temper the glass product.   Tempering is a key process in producing both automotive and architectural glass.   Conventionally, electric radiant heat has been used for glass tempering.   More recently, gas fired convection heat has been tested, with excellent results.

Industrial drying and curing

Many manufactured products require drying, including paper, paint, textiles, plastics, and even some fruits.   Electric infrared heat is used in many of these processes, but gas infrared burners have recently begun to compete effectively for this market. In the paper industry, ceramic and metallic gas infrared burners are used to dry coatings on paper and cardboard. More recently, these burners are being installed along the paper mill production line to boost drying capacity and overcome production bottlenecks. By pre-heating the wet paper before it enters the conventional dryer, gas infrared burners improve productivity, increase the dryer's speed and capacity, and increase process efficiency. Also, innovative types of gas infrared dryers and heaters are being developed for paper manufacturing applications.

Gas infrared heat is also used in automotive manufacturing to cure the paint on vehicles. Automotive paint is increasingly being applied as a powder instead of a solvent-based liquid. Powder paint eliminates the release of volatile chemicals, but requires precise control of time and temperature to ensure flawless results. Gas infrared burners can "gel" the powder paint in place before the vehicle enters the curing oven. Preheating with gas also increases productivity by allowing the vehicles to move through the oven more quickly.

In the plastics industry, electric dryers have been used almost exclusively to dry plastic resin before it was molded or extruded into finished products such as nylon and polyester. Recently, the gas industry developed and introduced industrial dryers designed for processing plastic resin materials. These plastic dryers can treat resins that are prone to absorb moisture (hygroscopic) or non-hygroscopic resins. They can also be used to modify the molecular structure of polyester before it is formed into beverage containers or other products.

Boilers and steam production

Manufacturers use steam heat for a multitude of processes in several industries. Large volumes of steam are required to produce chemicals, paper, and pharmaceuticals, for example. Process steam is generated ("raised") in large industrial boilers fueled by natural gas, oil, coal, wood waste (tree bark), or a mixture of fuels. As with other industrial processes, rising energy costs have driven improvements in boiler efficiency, and air quality regulations are beginning to require lower boiler emissions as well.

Boiler manufacturers have responded to these needs by developing and introducing natural gas-fired boilers with high efficiencies and very low emissions of nitrogen oxides (less than 25 parts per million). These industrial boilers are available in capacities up to 3,000 horsepower (2,240 kilowatts). Emissions of most air pollutants can also be reduced by using natural gas as a supplemental boiler fuel with coal or other primary fuels.

Food processing

The food industry is one of the largest manufacturing sectors in the U.S. and one of the biggest energy users. Most of the gas consumed in food processing is used in boilers to produce process steam, which is required for pasteurization, sterilization, canning, cooking, drying, packaging, equipment clean-up, and other processes. Food manufacturers were one of the first groups of energy consumers to install high-efficiency, low-emission natural gas-fired boilers.

The food processing industry also uses huge amounts of hot water for cleaning, blanching, bleaching, soaking, and sterilization. Natural gas is one of the industry's main fuels for heating water and other liquids. Special, very high-efficiency industrial water heaters have been developed for food and other applications (Fig. 7–14). In addition, food processing plants use gas for drying, cooking, and baking, as well as for refrigeration, freezing, and dehumidification.

Fig. 7–14 Direct Contact Water Heater (Courtesy of QuikWater)

Cogeneration

With boilers, the industrial use of natural gas crosses over into the realm of power generation, since the boiler's process steam can also be used to produce electricity. Instead of turning the boiler off when the steam is not needed for manufacturing, the boiler can keep running, and the steam can be diverted to a turbine for generating power. Some large industrial boilers can generate several thousands of kilowatts (megawatts) of power.

During the 1980s, federal regulations encouraged this practice by forcing local utilities to buy power from their customers, and many manufacturers installed power generation equipment "onsite," at the factory or mill. These facilities are also sometimes called "customer-sited" or "distributed" generation plants. If electricity and heat (steam or hot water) are both being produced at the same time, the term "cogeneration" applies. Cogeneration is more efficient than power generation alone because waste heat is recovered and used. Cogeneration is also called "combined heat and power" and, in the 1970s, used to be called a "total energy system."

Large industrial gas customers pioneered the use of onsite generation and cogeneration, but more recently, the gas industry has developed smaller equipment for small industrial and commercial customers as well. Typically, these systems produce less than 50 megawatts (MW) of power. In smaller cogeneration systems, a reciprocating engine can be used instead of a turbine to produce electricity, and heat is recovered from the engine's exhaust and radiator fluid. These systems are available in capacities of less than 5 MW.

Usually, these smaller commercial applications are economically feasible only if the customer has a practical use for the recovered heat, such as a swimming pool, laundry, or domestic hot water, or for heating and air conditioning a large building. (Recovered heat is suitable for powering absorption chillers.) In these cases, the overall efficiency of the cogeneration system can exceed 70%, well above that of conventional energy-using equipment. Although small cogeneration systems have had little success on the market, deregulation of the electric utility industry could open up opportunities for commercial and small industrial gas customers to produce their own power in distributed plants or cogeneration systems.

Power Generation

Traditionally, in addition to operating nuclear plants, utilities generate electricity in huge boilers that produce hundreds of megawatts of power. These base load utility power plants, which provide the bulk of the nation's electricity, are fueled mainly by coal. Smaller peakshaving plants, which supply electricity during the summer or other periods of high demand, are often powered by natural gas. Recently, however, large base load power plants have begun to use more natural gas to reduce emissions of sulfur dioxide, nitrogen oxides, and particulates (soot or smoke).

Gas turbines

Turbines, similar to a jet aircraft engine, can be powered by a variety of energy sources including hydropower (dams), nuclear power, or fossil fuels such as gas, oil, and coal. Turbines produce mechanical power to drive generators, which convert the turbine's energy to electricity. Natural gas can be used to produce electricity either directly, in a gas-powered turbine, or indirectly, in a steam-powered turbine (using steam from a gas-fired boiler). Many peakshaving plants use gas turbines to generate electricity.

Gas turbines can reach very high efficiencies (approaching 40%) and produce very low emissions of nitrogen oxides (below 25 parts per million) and other air pollutants. Some power plants use a "combined cycle," which incorporates both steam and gas turbines to reach higher overall efficiencies. In the U.S., the federal government supports research to develop more efficient, lower emission gas turbines, primarily to promote better air quality.

Cofiring and reburning

Base load utility power plants (also called central stations) are being required by federal regulations to reduce emissions of air pollutants. Sulfur dioxide emissions can be controlled by devices called electrostatic precipitators ("scrubbers"), and special burners and catalysts can reduce emissions of nitrogen oxides. However, these technologies can cause operating problems and, in some cases, can reduce the boiler's capacity, or output. Currently, many utilities and other power plant operators are turning to natural gas as a supplemental boiler fuel to reduce emissions and, as an added advantage, to boost boiler capacity. In these applications, gas contributes about 20% or less of the total heat input to the boiler; coal or other fuel remains the primary source of heat.

Cofiring is the simplest way to use gas in large utility boilers (Fig. 7–15). Natural gas burners are installed to inject gas into the boiler's combustion zone. The gas and coal "co-fire," or burn together. Cofiring reduces emissions of sulfur dioxide substantially, trims emissions of nitrogen oxides, and reduces "opacity" (clouds of particulate emissions).

In reburning, natural gas is injected above the boiler's coal burners to "re-burn" the products of combustion. Gas reburning can reduce nitrogen oxides emissions dramatically, by more than 50%. Both cofiring and reburning can be combined with conventional emission control technologies for further reductions.

Natural Gas as a Transportation Fuel

Gas demand in the transportation sector includes gas consumed by pipelines at compressor stations and natural gas used in vehicles. Compressor fuel accounts for only about 3% of total gas consumption. The amount of gas used as a vehicle fuel is negligible but could grow rapidly.

Natural gas vehicles

Natural gas can serve as a substitute for gasoline or diesel fuel in cars, trucks, buses, and other vehicles. Natural gas greatly reduces emissions of the air pollutants that cause urban smog. Besides improving air quality, natural gas can reduce

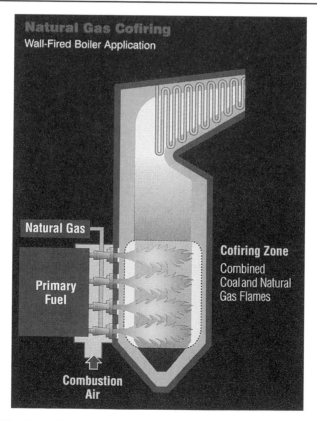

Fig. 7–15 Natural Gas Cofiring (Courtesy of Gas Research Institute)

fuel costs and increase engine life. Usually, the gas is compressed and stored in cylindrical tanks on-board the vehicle. However, some large, heavy-duty vehicles such as garbage trucks run on liquefied natural gas. In general, the use of natural gas as a vehicle fuel is practical only for fleets that are refueled at a central location, rather than in individual cars and trucks.

In the U.S., gas distribution companies pioneered the use of natural gas in their own fleets of vehicles. In other countries where gasoline and diesel fuel are much more expensive, or where air pollution is severe, governments have required conversion of vehicles to cleaner alternative fuels that are available domestically.

To improve air quality, U.S. federal regulations required fleets to begin using natural gas or other alternative transportation fuels beginning in the early 1990s. The U.S. Postal Service operates the largest fleet of alternative fuel vehicles in the country. In addition to federal law, most regions with large cities have their own vehicle emission standards. As of the late 1990s, tens of thousands of fleet vehicles in America were running on natural gas fuel, including urban transit buses, shuttle buses (especially at airports), school buses (Fig. 7–16), package delivery trucks, garbage trucks, street sweepers, forklift trucks, taxicabs, and even some police cars.

Fig. 7–16   **Natural Gas Fueled School Bus (Courtesy of Gas Research Institute)**

Natural gas fueled vehicles ("NGVs") can be either bi-fuel (dual-fuel) or dedicated. Bi-fuel vehicles can run on either natural gas or conventional fuel (typically gasoline); they have a switch for the driver to revert to gasoline when the natural gas runs out. Many bi-fuel vehicles were converted with a relatively simple kit of equipment, either in the factory or by the fleet owner. Dedicated NGVs carry only natural gas and have a limited traveling distance (less than 200 miles per fill-up).

The major U.S. automakers, and foreign manufacturers as well, offer bi-fuel natural gas/gasoline cars, vans, and trucks. Ford sells dedicated NGV cars, vans, and trucks, and DaimlerChrysler offers a dedicated NGV van. Most manufacturers of heavy-duty engines offer dedicated natural gas products for buses and trucks. Often, these large engines can run on either compressed or liquefied natural gas.

The market for NGVs in the U.S. has been limited by the lack of stations where drivers can fill up their tanks. Less than 1,500 natural gas refueling stations were in operation in 1997. Fleet owners usually install private stations to serve their own vehicles. Although more public stations are being built, they are still not widespread enough to make NGVs convenient for everyday use. Also, NGVs still cost substantially more than conventional vehicles. The gas industry is developing fuel storage technology to increase the traveling range of NGVs and reduce their price tag.

Bibliography

Gas Appliance Technology Center, *GATC Focus*.

Gas Research Institute Digest.

American Gas Association, *Gas Technology*.

8

REGULATORY HISTORY OF THE GAS INDUSTRY

Introduction

For our purposes in this chapter, regulation is defined as a rule or order having the force of law issued by an executive authority of a government. Governing regulations of one sort or another have been imposed since the dawn of civilization. In modern times, regulations restrict the activities of individuals and corporations in the interest of the health, safety, and general welfare of the public. Regulations are based on laws, or legislation, enacted by governments in accordance with the "police powers" granted by a constitution or other mechanism on which the government is founded.

In the United States, the Constitution was written at a time when economists such as Adam Smith were espousing the concept of *laissez-faire* (loosely translated, "let it be"). These natural laws of production and exchanges, whose principles were incorporated into the Constitution, assumed that a market would be self-regulating – that is, a producer would set the price for its product at what the market would bear. If the product was overpriced, consumers would simply do without it or use an equivalent, lower priced product.

Important U.S. Supreme Court cases have affected the ability of government to regulate businesses by defining a company as an individual. Since these early rulings, the courts have rendered many more decisions that recognize the right of government to control the activities of public utilities and other companies through regulation.

Utility regulation is based on the premise that a natural monopoly, franchised to be the exclusive supplier of a utility service to a specific area, is preferable to allowing multiple companies to compete for consumers, which is considered wasteful. As an alternative to control through competition, control is imposed through regulation to ensure adequate service at fair and equitable rates.

The basic mission of a public utility regulatory agency is

1. To ensure consumers that utility service meets prescribed standards, the quality of service is commensurate with the charges for that service, and the service will be adequate to meet demand over the long term
2. To allow investors an opportunity to receive a fair and reasonable return on their investment

The Early Days of Regulation

The early gas companies needed large capital investments to enable them to construct plants for manufacturing gas and to lay pipes in the streets. Investment in gas companies was considered risky. To overcome this concern, the companies obtained exclusive franchises to provide service in an area. A franchise gave the company the right to install, operate, and maintain its facilities on public streets and highways. An exclusive franchise ensured that no other gas company would be permitted these same rights in the same area. This reduced the investment risk, and gas companies were able to obtain the capital required for their facilities.

The gas industry grew rapidly during the mid-1800s. Gas companies were recognized as good investment opportunities, and the exclusive franchise was no longer an important factor for raising capital. By late in the century, its use had all but disappeared. This led to intense competition, as gas companies installed their facilities side-by-side with those of other gas companies. In many cities, price wars raged until only one company remained. This company would then raise its rates to exorbitant levels to recover its losses.

Governmental bodies recognized that unrestrained competition between companies providing the same utility service was wasteful and that something had to be done to control these activities. The solution was to develop municipal ownership of the utility service or to impose regulations on privately owned utilities.

At about this time, the gas industry faced a new competitive challenge: the electric light, invented in 1876. The newly formed electric utilities quickly penetrated the lighting market. As a result, many of the smaller gas utilities that could not withstand the competitive onslaught were merged or consolidated into large gas utilities. With a stronger financial base, these larger companies were able to develop improved processes for manufacturing gas and to promote new uses for gas such as cooking, water heating, and space heating. With a competitive edge, gas lighting remained an important load for gas utilities through the early 1900s.

Early attempts at regulation of gas utilities by state governments were generally unsuccessful. Many states had created Railroad Commissions to regulate the activities of that rapidly expanding industry. In some states, these commissions were empowered to regulate public utilities as well. The effectiveness of these commissions was questionable, with most merely making observations and recommendations. The first state agency with power to regulate gas and electric companies by mandatory imposition of rules and regulations in the public interest was the Massachusetts Board of Gas Commissioners, established in 1885.

Comprehensive regulatory statutes creating public utility commissions were first enacted in Wisconsin and New York in 1907. Other states quickly followed suit, and as of the 1990s, all 50 states have commissions that regulate activities of stockholder-owned public utilities, which include gas distribution companies. On the federal level, enactment of the Interstate Commerce Act in 1887 affected the operations of companies transporting goods and services across state lines, but an amendment to this act in 1906 specifically excluded natural gas pipeline companies from the jurisdiction of the Interstate Commerce Commission.

From 1900 to 1930, several holding companies were formed to manage the operations of emerging natural gas pipeline companies. Many of these companies also controlled gas distribution companies in several different states. The Public Utility Holding Company Act of 1935 gave the Securities and Exchange Commission extensive jurisdiction over electric and gas holding companies. This authority applied to any company that exercised control over 10% or more of the voting securities of a company. As a result of this act, many gas distribution companies were spun off to become independently owned and operated.

The Public Utility Holding Company Act did not apply, however, to companies controlling interstate transmission pipelines. As it became practical to transport natural gas long distances from production fields to industrial areas, the facilities of many of the large transmission companies traversed several states. However, state regulators did not have the authority to control the rates or conditions of service imposed by interstate pipeline companies.

At the end of World War II, the gas industry experienced a period of rapid growth. The availability of natural gas to the Northeast, along with growth in the nation's economy, resulted in huge increases in demand for natural gas service.

In fact, demand was so great that, at times, it could not be fully met until additional pipeline facilities were completed and placed in service. This rapid expansion was also a period of economic inflation, so that gas utilities had to file frequently for rate increases to offset increases in operating expenses.

Federal Regulation

Like all industries in America, the gas industry is subject to many federal regulations promulgated by agencies or commissions that were created by federal legislative acts. Examples include the Occupational Health and Safety Administration, which regulates employee safety, and the Securities Exchange Commission, which regulates financing and financial reporting. In addition to these general industry regulations, federal agencies have been created by law with responsibilities specifically related to the energy industry, which includes natural gas transmission and distribution companies. Table 8–1 summarizes federal legislation and other actions affecting the natural gas industry.

The first federal act directly affecting the gas industry was the Natural Gas Act of 1938. This authorized the Federal Power Commission (FPC), which had been created in 1920 to regulate construction of hydroelectric dams, to regulate activities of interstate pipeline companies. The FPC was given specific authority to:

- Regulate the flow of natural gas in interstate commerce
- Set rates and tariffs and approve contracts for the sale and transportation of natural gas
- Grant certificates of public convenience and necessity to interstate pipeline companies and require these companies to provide natural gas services to municipalities or corporations engaged in the business of local gas distribution
- Provide a uniform system of accounts for interstate pipeline companies
- Require periodic and special reports as necessary for the administration of its activities
- Aid in the conservation of natural gas resources

When it was passed, the Natural Gas Act was assumed not to apply to gas producers. Following World War II, however, with inflation and the ensuing sudden increase in gas demand, the price that producers charged for gas was raised substantially. These increases concerned the pipeline and distribution companies, which were required to apply for rate increases to pass on these higher costs to their customers. This situation changed in 1954 when, in the Phillips Petroleum

Table 8–1. Federal Legislation, Regulation, and Other Actions Affecting the Natural Gas Industry

Action	Year	Summary
Natural Gas Act	1938	Authorized the Federal Power Commission (FPC) to regulate interstate pipeline companies
U.S. Supreme Court ruling in Phillips PetroleumCompany Case	1954	Gave the FPC authority to regulate the price that producers charge for natural gas
Federal Energy Administration Act	1974	Gave the administration power to allocate and control pricing of scarce petroleum products including gas
Department of Energy (DOE) Organization Act	1974	Created the DOE and the Federal Energy Regulatory Commission (FERC)
National Energy Act (five parts) National Energy Conservation Policy Act	1978	Encouraged utilities and their customers to conserve energy
Power Plant and Industrial Fuel Use Act		Prohibited the use of natural gas in utility and industrial boilers
Public Utility Regulatory Policies Act		Encouraged cogeneration of heat and power by industrial customers
Natural Gas Policy Act		Gave producers more incentive by phasing out regulation of gas prices at the wellhead
Energy Tax Act		Penalized low-mileage autos and rewarded conservation measures
Natural Gas Wellhead Decontrol Act	1989	Completed deregulation of wellhead gas prices
Federal Energy Regulatory Commission Orders (436, 500, 636)	1985-1993	Deregulated pipeline transportation, allowing customers to buy gas directly
Clean Air Act Amendments	1990	Empowered the Environmental Protection Agency to set national air quality standards to curb acid rain, urban pollution, and toxic emissions
Energy Policy Act	1992	Mandated purchase of alternative fuel fleet vehicles to reduce America's dependence on foreign oil

Case, the U.S. Supreme Court ruled that producers were indeed covered by the Natural Gas Act and were therefore under the jurisdiction of the FPC.

Following this decision, a rash of rate applications by producers overwhelmed the administrative capacity of the FPC. In an attempt to control the number of applications, it implemented a rule to establish pricing by production area. These pricing regulations proved to be a disincentive for producers, who found that exploration for oil, particularly in foreign countries, was far more lucrative than developing natural gas reserves in the United States. As a result, U.S. gas drilling activity dropped off and, by the mid-1960s, some pipeline companies were unable to purchase the gas volumes necessary to meet their existing sales commitments.

"Curtailment," referring to gas supply cut-off, became the by-word of the day in the gas industry. The FPC established a priority system that distinguished "human-needs" customers (the group most in need of a guaranteed supply) from "interruptible" consumers, such as power plants that could use other fuels. These interruptible users became the first candidates for curtailment when gas supply could not meet demand.

Then in 1973, the Organization of Petroleum Exporting Countries' oil embargo led to passage of the Federal Energy Administration Act the following year. The Federal Energy Administration was empowered to allocate and control pricing of scarce petroleum products, including natural gas. Around the same time, the Department of Energy Organization Act created the Federal Energy Regulatory Commission, which assumed many of the FPC's responsibilities.

National Energy Act

In 1978, a legislative landmark – the National Energy Act – was passed, which was comprised of five separate acts:

- *National Energy Conservation Policy Act* required utilities to encourage their customers to conserve energy, including the financing of conservation measures through utility bills.
- *Power Plant and Industrial Fuel Use Act* required power plants to convert to coal, where feasible; in effect, it prohibited the use of gas in utility and industrial boilers.
- *Public Utility Regulatory Policies Act* established federal standards for termination of service, required that utility stockholders pay for advertising of a political or promotional nature, and spurred development of cogeneration projects.
- *Natural Gas Policy Act* provided for the gradual phase-out of producer rate regulation, redefined curtailment priorities for agricultural use of natural

gas, and provided incremental pricing for industrial sales (lower rates based on supplying just the additional amount, or next increment, of gas).

• *Energy Tax Act* established tax credits for dwelling units that implemented conservation measures and imposed an excise tax on automobiles with low gasoline mileage.

The National Energy Act transformed the gas industry and initiated a supply-demand roller coaster ride. Before the act, many new gas discoveries were dedicated to the intrastate market, where producers could get higher prices not subject to federal controls. Afterwards, interstate pipeline companies were able to compete with their intrastate counterparts by offering to purchase gas at the same higher prices as intrastate buyers.

Over the next few years, the gas supply situation reversed. As prices increased, large volumes of gas came onto the market, and supply soon exceeded demand, resulting in a gas "bubble" of over-supply. At the same time, however, higher prices for gas eroded its competitive position with oil, and large industrial customers switched from gas to other energy sources. As industrial gas demand dropped off, pipeline companies were unable to "take," or accept, their full contractual quantities of gas from producers.

Where possible, the pipelines exercised legal "market-out" options, also called *force majeure*. These options allowed companies to prove in court that the extremely unusual market conditions were similar to a natural disaster. Otherwise, given the "take-or-pay" provisions of their contracts, the pipelines were required to pay for the gas even if they couldn't take it and sell it.

Take-or-pay provisions were meant to ensure gas suppliers that all their production costs would be recovered. The Federal Energy Regulatory Commission (FERC) allowed part of these costs to be passed on to local gas distribution companies. The state public utility commissions were left with the job of determining the method by which the distribution companies in their jurisdiction could recover these costs.

One section of the 1978 National Energy Act, the Public Utility Regulatory Policies Act (PURPA), had a profound impact on the economics of cogeneration, defined as the simultaneous production of heat and electricity (see Chapter 7, *Uses for Natural Gas*). Cogeneration is more efficient than power generation alone because waste heat is recovered and used. Previously, industrial gas customers who wanted to use excess steam to generate electricity were required to connect with the utility grid, an expensive and complex procedure. Also, electric utilities charged high rates for backup power to these customers.

PURPA eliminated the grid connection requirement, obliged utilities to provide backup power at reasonable rates, and went even further, requiring them to buy power from their industrial customers at the "avoided cost" (whatever it would cost the utility to generate the electricity itself). By essentially guaranteeing a price and market for cogenerated power, PURPA greatly improved the economic attractiveness of cogeneration for many industrial gas customers. It also spawned competition against utilities in the form of "independent power producers," who built new plants solely for cogeneration purposes. Utilities challenged PURPA in court, but its rules were implemented in 1983.

All remaining price controls on natural gas at the wellhead were terminated by the end of 1992 under provisions of the Natural Gas Wellhead Decontrol Act of 1989. Congress recognized that the thousands of producers and many buyers involved in the gas market already made it highly competitive. On a federal level, this act created a free market climate for sale of natural gas, resulting in much lower prices on the short-term "spot" market.

Deregulation of gas transportation

Perhaps the greatest impact of federal regulation came as a result of FERC orders implementing the Natural Gas Policy Act (Order No. 436 in 1985, No. 500 in 1987, and No. 636 in 1993). These orders permitted local gas distributors and large customers to "by-pass" the pipeline and purchase gas directly from producers, marketers, and brokers. For a while, some customers were building their own pipelines, but FERC soon required pipeline companies to transport any purchased gas, resulting in a radical change in the seller-purchaser relationship of the past. State public utility commissions then had to consider the impact of by-pass on the companies under their jurisdiction.

By 1993, FERC orders had provided for:

- Fully comparable transportation services for gas, whether sold by the pipeline or by a third party
- Open-access storage service
- Separation of purchase and transportation services by interstate pipelines ("unbundling")
- Deregulation of interstate pipeline sales sources, with rates for sales of natural gas becoming constrained only by the market
- Pre-granted abandonment rules, allowing gas sales contracts to terminate at the date specified in the contract, with no rules or provisions for continuation of service
- Rates for transportation and storage service with all fixed costs, including return on equity and associated income taxes, to be included in demand charges
- All information and discounts available to company marketing affiliates be available to all entities
- Agreement that the transition costs associated with compliance would be considered in the pipeline company's next general rate filing

Environmental regulation

The original Clean Air Act was passed in 1970, but Congress approved a series of sweeping revisions in 1990 to enforce the law and give it some real teeth. These amendments are intended to curb acid rain, urban air pollution, and toxic air emissions (hazardous pollutants). The natural gas industry is not a specific target of the law, but has benefited from some of its provisions while having to comply with others.

For example, power plants produce emissions of sulfur dioxide and nitrogen oxides, which both contribute to acid rain. As a result of gradual enforcement of the Clean Air Act Amendments, power plant operators have begun to use more natural gas to reduce emissions by cofiring and reburning (see chapter 7, *Uses for Natural Gas*). Industrial energy consumers have also turned increasingly to natural gas fuel for boilers and have developed innovative operating strategies for gas-fueled furnaces and other equipment.

Similarly, cars and other vehicles running on gasoline and diesel fuel emit chemicals that react with ozone to cause urban smog, which is the most prevalent air pollution problem in the U.S. More than 20 cities have not yet complied with the federal air quality standard for ozone. In these areas, many natural gas vehicles have been purchased to meet local and regional air quality standards, which are required by the Clean Air Act Amendments. When used in vehicles, natural gas produces lower emissions of many pollutants than gasoline or diesel fuel.

On the other hand, gas-burning equipment must also comply with standards initiated by the Clean Air Act Amendments. These have affected numerous gas-fired engines and turbines, especially those in service at pipeline compressor stations. Technologies continue to be developed to reduce emissions from these engines and turbines. In southern California, even relatively small engines, such as those used in gas-powered chillers, must be equipped with pollution control devices to obtain an operating permit. The gas processing industry is also being affected by the amendments. Dehydration plants, which use glycol to remove water from raw natural gas, emit toxic air pollutants. Gas processors are currently evaluating ways to control emissions of these chemicals.

Another environmental concern is depletion of the earth's ozone layer, an atmospheric screen that protects people from ultraviolet rays. Chemicals containing chlorine gas, such as chlorofluorocarbon refrigerants used in electric chillers and automotive air conditioners, are thought to be the main culprit. During the 1990s, these chemicals were phased out by the Environmental Protection Agency, and many electric chillers are being replaced with gas-powered systems, which use no harmful refrigerants.

An environmental controversy being discussed worldwide during the 1990s is global warming. This is thought to be caused by "greenhouse" gases such as carbon dioxide and methane, which trap the earth's heat and gradually warm the atmosphere. Warmer air and water temperatures can disrupt weather patterns and could cause major changes in the world's climates.

The gas industry might have to develop ways to control emissions of methane from vehicles, pipelines, and other facilities. However, natural gas can produce lower emissions of carbon dioxide than other fossil fuels when used in high-efficiency turbines, boilers, and other equipment, leading to overall reduction of global warming.

Energy Policy Act

The Energy Policy Act of 1992 was another landmark piece of legislation, in terms of its effect on the natural gas industry. This law was designed to reduce America's dependence on foreign oil. It requires the federal government to buy vehicles that can run on alternative fuels for a certain percentage of its fleets and to provide incentives for the private sector to do the same. The law, which applies to 125 metropolitan areas (regardless of their air quality), mandates incorporation of alternative fuel vehicles into certain centrally fueled fleets and provides tax deductions for vehicles equipped to operate on an alternative fuel. Many natural gas vehicles have been purchased by fleets, both government and private, as a result of these provisions.

Other sections of the Energy Policy Act authorize federal funding of research to recover more natural gas from conventional and unconventional sources and to improve the nation's gas storage capability. It also authorizes spending to perfect new gas technologies that reduce emissions from stationary sources, such as power plants, and to develop energy-efficient heating and cooling technologies for residential and commercial buildings. The act also makes it easier for non-utility and unregulated utility spin-offs to build electric generating plants and to gain access to the nationwide utility-owned electric transmission grid.

The act also covers energy efficiency standards for appliances and other equipment. It requires state regulators to consider "integrated resource" plans for gas utilities. These establish supply and demand strategies that are supposed to minimize costs and increase energy efficiency. State regulators must also consider allowing utilities to earn a profit on their investments in techniques that improve energy efficiency.

State Regulatory Commissions

As mentioned above, all 50 states have enacted laws establishing regulatory agencies that are empowered with jurisdiction over gas distribution companies. The common names used are Public Service Commission and Public Utility Commission; in Texas, jurisdiction over utilities resides with the Railroad Commission. In general, regulatory authority is vested with these commissions over privately owned gas distribution and intrastate pipeline companies. In some states, the commissions also have jurisdiction over publicly owned utilities and, if certified by the U.S. Department of Transportation, they are responsible for enforcing safety regulations for all gas operations, regardless of ownership.

With few exceptions and minor differences, state utility commissions have the authority to regulate gas distribution company functions relating to:

- General terms and conditions of service
- Rates and tariffs
- Financing
- Accounting and reporting requirements
- Mergers and acquisitions
- Purchase and disposal of facilities
- Construction and extension of facilities

Unusual situations

Although much public attention is drawn to rate proceedings that have a direct impact on customers, public utility commissions have affected gas distribution operations in a number of other ways. Because gas distribution is a dynamic industry, situations arise from time to time that require a decision by the regulatory authority. For example:

- In the late 1930s, commission activity concentrated on adoption of a uniform system of accounts for public utilities.
- During the 1940s - 1960s, regulatory agencies had to cope with the accounting treatment for recovery of utility costs incurred when converting from manufactured to natural gas.
- In the 1960s, during a period of intense competition between gas companies, electric utilities, and unregulated oil dealers, commissions had to consider allowances for promotional activities.
- The 1970s saw the introduction of utility management audits to ascertain whether improvements should be made to increase efficiency and reduce operating costs.

Commissions in the 1970s also were required to devise a way for gas companies to recover their cost of gas, which increased rapidly over a short period of time. The conventional rate case application and hearing process did not effectively provide cost relief. To enable gas companies to recover these costs more quickly, "purchased gas adjustment" clauses were introduced. Commissions continue to review the prudence of gas purchase practices to ensure that the company is acquiring gas supplies at the lowest possible cost.

Utility rate applications

Public utility commissions have established procedures for rate cases. Not many people realize that utilities must apply to reduce their rates as well as to increase them. In its application, the gas company submits documentation to support its request in a format prescribed by the commission. These documents are reviewed by commission staff, and a formal hearing is conducted before one of the commissioners (acting as an examiner) or before an administrative law judge appointed to hear testimony. These are public hearings, and intervention by customers and organizations representing customers is permitted. Most states have established a consumer advocate's office, which intervenes and participates in the hearings on behalf of customers.

An initial decision is rendered by the hearing examiner for review by the full commission, which issues a final decision. The decision states the just and reasonable rate of return to which the company is entitled and, based on financial data, the additional revenues that the company will require to achieve this rate of return. The order instructs the company to file rate schedules by class of service that should achieve these revenues.

Following another review, the commission approves the rate schedules. Most commissions require a cost of service study that identifies the revenues required from each class of customers based on the cost of providing the service to that class. An appeals process is usually established for cases in which a company feels that the commission's decision is unfair. After exhausting administrative appeals, the company has the right to seek recourse in the courts.

Just because the commission establishes an allowed rate of return, however, does not guarantee that the company will attain that return. Any number of factors, such as loss of a large customer, warmer than normal weather, or a sudden increase in the cost of gas can reduce revenues and increase operating expenses, preventing the company from attaining the return determined by the commission as just and reasonable.

On the other hand, in rare cases when a company consistently exceeds its allowed rate of return, the commission is empowered to hear evidence showing why the company should not reduce its rates to provide only those revenues necessary to produce the allowed rate of return.

Other routine activities

In addition to hearing rate cases and resolving unusual problems, commissions perform many other functions on a continuing, day-to-day basis, such as determining the standards of service. The commission prescribes an allowable pressure range for gas deliveries to the customer that assures proper operation of gas equipment. Commissions also set standards for the accuracy of meters that measure the volumes of gas delivered to customers and procedures for periodic testing of these meters to determine that they have functioned accurately. Commissions also may prescribe the method of billing for the service, whether by volume or by heating value (therms or Btus). Other billing matters relate to the frequency of meter reading and billing and the termination of service for nonpayment.

State regulatory commissions also prevent undue or unreasonable discrimination of a customer or class of customers. Discrimination exists not only when customers with identical requirements are treated differently, but also when a customers with dissimilar service characteristics or requirements are treated alike. To avoid discrimination, state commissions have the authority to determine the level of rates to each class of customer.

Other matters that arise – most frequently during a rate case, but also as a result of a customer complaint – relate to various discretionary operating expenses allocated in determining the cost of service. Issues relating to service on a customer's premises, such as appliance adjustment and repair, might be considered discriminatory because not all customers share equally in the service offered. Many commissions require companies to charge customers for these services as they are rendered.

Advertising and promotional expenses are also the subject of commissions' scrutiny and might not be allowed as a cost of service unless the company can show that these expenses benefit all customers. Treating charitable contributions as an operating expense has also been contested in some states.

Municipal Regulations

Gas distribution companies are also subject to regulation by the municipal governments in their service areas. Franchises granted by municipalities generally specify a time period in which the franchise is effective, after which it will expire unless the utility applies for an extension of the franchise privilege. Municipalities also impose a franchise tax on the utility, which is usually determined as a percent of the revenue received from utility sales within the municipality.

Although the franchise gives the utility the right to install, operate, and maintain its facilities in public rights of way, the municipality usually requires the utility to obtain a permit for any excavation work and pay a fee for inspection of the excavation to ensure that the surface is restored to its original or better condition. Gas distribution companies must also comply with municipal regulations pertaining to fuel line piping, gas appliances, and gas equipment.

Safety Regulations

Throughout its early years, the gas industry's public safety programs were self-policing. Considering the explosive and toxic nature of manufactured gas, gas distribution companies used extreme care in their operations to ensure public safety. The American Standards Association published the first code for pressure piping in 1942. A separate section dealing exclusively with gas transmission and distribution pipelines was issued in 1952 and has been updated regularly since then. The committee responsible for maintaining gas piping standards resides with the American Gas Association.

The Natural Gas Pipeline Safety Act of 1968 was the first national regulation for safety of gas system operation, maintenance, design, and construction. It empowered the U.S. Department of Transportation to prescribe minimum federal safety standards for all gas operations. These regulations also are updated regularly; as of January 1996, they had been amended 74 times. Administration and enforcement of pipeline safety regulations can be delegated to certified state agencies.

Safety standards are periodically reauthorized by passage of new legislation. For example, the Pipeline Safety Act of 1992 reauthorized the original Natural Gas Pipeline Safety Act and the Hazardous Liquid Pipeline Safety Act. Reauthorization establishes the ground rules and deadlines for these laws, but leaves detailed requirements to be determined by the Department of Transportation. As a notable change in the 1992 act, the Office of Pipeline Safety was required to consider environmental impacts as well as safety when developing regulations.

Public safety is also a primary concern of state regulatory commissions. Safety of the customer's fuel line piping and gas equipment is regulated by municipal codes that usually use the National Fuel Gas Code as a standard. State commissions are responsible for rules pertaining to termination of service in the event of a problem with a customer's piping or equipment.

Gas Industry Organizations

The U.S. natural gas industry and its counterpart in Canada support a variety of organizations that work with regulators to promote safety, maintain codes and standards, develop new technologies, and disseminate information. These organizations and their memberships include:

- American Gas Association (local distribution companies)
- American Public Gas Association (municipal utilities)
- Canadian Gas Association (gas utilities)
- Canadian Gas Research Institute (membership organization, industry-wide)
- Gas Appliance Manufacturers Association
- Gas Research Institute (not-for-profit membership organization, industry-wide)
- Institute of Gas Technology (not-for-profit energy education and research, industry-wide)
- Interstate Natural Gas Association of America (pipeline companies)
- Natural Gas Supply Association (producers)

In addition, various trade groups work with regulators in specific areas of the natural gas industry, such as:

- American Gas Cooling Center
- Industrial Center
- Natural Gas Vehicle Coalition

9

GAS MARKETING AND SALES

Introduction

Until recently, no one would have predicted that natural gas would be sold door-to-door along with cosmetics and kitchen cleansers. Yet in 1998, Columbia Energy Services announced a partnership with Amway Corporation to do just that. Amway's representatives will sell gas service to homeowners and small business operators wherever individual states allow such retail sales. The idea is to introduce competition, give consumers a choice, and let them shop for energy providers the way they do for long-distance telephone service.

Still, most natural gas is marketed in conventional ways. Marketing is a term that is often misunderstood to mean merely selling, when actually it involves researching consumer needs and devising ways to fulfill consumer demand. Effective gas marketing ultimately results in a sale (purchase contract), subsequent transportation of gas, and delivery of the product or service.

Since gas customers began buying directly from producers and other suppliers, natural gas has become a commodity like grain or pork bellies. But that doesn't mean that the gas industry has stopped marketing its product. In fact, natural gas marketing is a growing industry, with producers, pipeline companies, independent traders and marketers, and traditional gas utilities all vying for a piece of the action.

Brief History of Gas Marketing

Until the 1980s, federal and state agencies regulated the rates that pipelines and distribution companies could charge their customers and controlled the price of natural gas at the wellhead (see Chapter 8, *Regulatory History of the Gas Industry*). Pipelines bought natural gas from producers, transported it to markets, then resold the gas to local distribution companies. Producers sold gas under long-term (typically, 20-year) contracts to pipelines, who agreed to pay for the gas whether they could take it and sell it or not ("take-or-pay" provisions).

Soon, pipelines found themselves with a surplus of gas they could not sell, but were still required to pay a high price for. In the mid-80s, the pipeline industry began a series of painful changes as federal agencies deregulated gas sales, allowing customers to buy gas directly from producers.

Pipeline companies were suddenly transformed from merchants to transporters. The opportunity for direct sales, along with the gas surplus that persisted through most of the decade, made competitive marketing the biggest challenge for pipelines and producers alike.

Pipeline Marketing and Transportation

One result of these regulatory changes was the proliferation of gas marketing companies, which are often subsidiaries spun off by pipelines, producers, and distribution companies. These marketers assemble gas from several sources into large-volume packages for sale to industrial users and local gas distribution companies. In addition, independent marketers (not affiliated with any gas company) have sprung up and captured a share of the gas marketing business. More than 300 gas marketers are in business, including both independent and affiliated companies.

Besides creating subsidiaries, many pipelines have reorganized by separating their marketing units from their transportation operations, and their merchant role has become much smaller. Pipeline marketing units seek to attract new customers and are driven by sales volume and profit margin. Operations divisions focus on reducing transportation costs, increasing reliability, and broadening the range of transportation services while keeping the pipeline as full of gas as possible. These competitive efforts have largely succeeded. In fact, due to more efficient, less expensive transportation, the price of gas is expected to fall in real terms over the next decade (see Chapter 10, *Future Supply and Demand for Natural Gas*).

Sales contracts

As a result of deregulation and industry restructuring, the length, or term, of gas purchase contracts has changed dramatically. In contrast to historical 20-year contracts, by the late 1980s most gas was sold through 30-day transactions. This short-term market for gas has grown rapidly and has become the source of large quantities of gas sales. To accommodate these transactions, pipelines offer shorter term transportation arrangements.

Recently, some gas sellers have begun to return to longer term contracts in order to gain customers' confidence in the natural gas supply and to capture new markets such as power generation plants. This required creating ways to accommodate changes in gas prices over the longer term. Purchase agreements have become a mixed bag of short-, mid-, and long-term contracts, depending on the needs of the customer.

The profound changes in the way that gas is bought, sold, and transported have created headaches for pipeline companies. Suddenly, they had many more players and transactions to deal with. Also, because many deals are short term, the parties involved change frequently, in contrast to the stable buyer-seller relationships of the past. Arranging and accounting for all these transactions, from wellhead to market, has become a critical function of pipelines.

Gas customers perform daily contractual operations called "nominations," which refer to the amount of gas to be delivered on a given day. For example, a customer might have a one-year contract with a supplier for a certain volume of gas, specifying an annual minimum and maximum amount. But each day by 8:00 a.m., the customer must tell the pipeline company how much of that gas to deliver during the next 24 hours and where to deliver it.

Nominations are required to keep the pipeline system balanced. Pipelines also make allocations among their customers, depending on whether the customer is firm or interruptible. Given a limited pipeline transportation capacity, firm customers are served first, then capacity is allocated to interruptible customers.

Gas Marketers and Traders

Since deregulation of wellhead prices and pipeline transport, natural gas has been traded as a commodity, or raw material, on financial exchanges. This means that in addition to the physical market for gas, where the product is actually transported and delivered, a financial market for gas has developed and begun to thrive. Financial players benefit the marketplace by spreading risk, stabilizing prices, and increasing liquidity (the flexibility of turning an asset into cash).

Independent marketers and gas company subsidiaries act as middlemen, buying gas from producers, arranging for transportation, and selling it to customers. Marketers actually own the gas (take title to it) for at least a brief period of time and usually make deals to supply gas for a year or more. In contrast, gas traders (known as "day traders" or brokers) are also middlemen, but never actually own the gas. Traders and brokers make very short term deals, usually only a few months and as little as one day.

Deals between gas marketers, as opposed to transactions between gas producers and marketers, have begun to account for a greater percentage of total transactions. In 1997 on the buy side, 47% of transactions were marketer-to-marketer deals, versus 50% with producers. On the sell side, 22% of transactions were marketers' sales to other marketers. Deals between marketers are important because they help create market liquidity.

Gas marketers and traders also make deals with their counterparts in the electric power industry, which began making the transition toward a commodity market during the 1990s. These gas/power transactions can be especially lucrative deals. For example, traders can take advantage of a spike in gas prices in New York by turning off a gas-fired power plant there and selling the excess gas at local market prices. To make up for the power plant being turned off, the trader could buy electric power in Ohio and transmit it to New York.

Spot market

In commodity transactions, traders buy and sell by paying the full cash price "on the spot" (thus the term, "spot market"). Commodity markets are risky for the investor, since natural disasters and global politics can spark huge price fluctuations, or volatility. Natural gas is a highly volatile commodity anyway, especially just before winter sets in, as traders make guesses about the severity of the cold weather to come. Spot market prices give gas consumers a clue as to what they can expect to be paying over the near term.

"Confirmations," another type of contractual operation, help traders manage the risk of their deals. Confirmations refer to an independent verification of a contract. After two traders do a deal, they hand the contract off to their "back offices" (accountants and other administrators), who communicate with each other to confirm and execute the deal.

Basis trading

The difference in the price of natural gas at one location on the pipeline and at another location is called the "market basis differential". A molecule of gas in different locations is worth different values, depending on the local supply/demand situation. In the financial gas market, deals are made through "basis trading" rather than by physically moving the gas. Traders maximize their profits by buying gas where it is cheap and selling it where it is expensive, without necessarily transporting any gas.

Pipeline companies can participate in this market to a certain extent, but regulatory tariffs limit how much of the difference in gas price between two locations that the pipeline can capture. Price differences up to the tariff rate go to the pipeline as transportation revenue. Beyond that, the excess goes to the gas trader or marketer.

Futures and options

Gas futures and options contracts are derivative investments that represent a bet on how natural gas prices will move up or down in the future. A futures contract obligates the investor to buy or sell a volume of gas on a specific day for a pre-determined price, while options give the investor the right to buy or sell a gas volume for a pre-determined price at any point over a specified period of time.

Futures and options can help buyers and sellers minimize the risk of rapidly rising or falling gas prices. For example, if a buyer has a futures contract to purchase a certain volume gas for $3.00 per unit later that year, the buyer is protected if the spot market price jumps to $4.00 per unit of gas before then.

Gas marketers hold more than half of the open interest in gas futures contracts on the New York Mercantile Exchange. This makes sense because marketers bear the most risk of price exposure on both the buy and sell sides of their business. The rest of futures contracts interest is held by producers, financiers, speculators, and others.

Distribution Company Marketing

In addition to these physical and financial gas markets, local gas distribution companies market gas primarily by promoting its use in general, as well as the use of gas-burning equipment through cooperative advertising with appliance dealers and manufacturers. Many distribution companies are no longer involved in direct merchandising of equipment, but their staffs assist engineers, architects, developers, plant managers, and commercial and industrial customers in equipment selection. Related marketing efforts include energy conservation programs and energy audits of homes and businesses.

However, some distribution companies do sell gas appliances and equipment directly to users, particularly in areas where local dealers do not promote gas aggressively. In fact, gas utilities are competing with dealers in certain high-growth markets, such as gas fireplaces, logs, and other hearth products.

These activities are regulated to prevent utilities from subsidizing their merchandising efforts with general funds collected from ratepayers.

Sometimes, local utilities also compete with contractors who install heating and air conditioning equipment. Utilities can own subsidiary contracting companies as long as their financing is kept separate from the utility's regulated business.

Current Gas Marketplace

Most American businesses are beginning to see the benefits of gas industry deregulation. Business customers can be divided into transport clients, who pay brokers or energy marketers for delivered gas, and sales customers, who depend on a single gas utility for energy supply and services.

About 60% of industrial customers and 40% of commercial customers have the option to buy gas from a supplier other than their local utility, according to a recent national survey. Soon, homeowners will have the same choice as retail residential gas sales are gradually deregulated. This has already begun in some test markets such as the Chicago area.

Most commercial and industrial customers are fairly satisfied with the services and information provided by their gas supplier, the survey showed. Overall, gas marketers received higher marks than full-service utilities. However, nearly a third of utility sales customers are former transport clients who returned to the local utility. Also, when customers have a choice between a marketer and a full-service utility, they tend to stay with the utility.

Despite fairly widespread satisfaction with marketers and utilities, however, more than half of all business customers have not seen a major reduction in their annual gas costs. A significant percentage of customers are disappointed with the performance of their suppliers (both marketers and utilities) in terms of cost containment, pricing options, and ease of doing business.

As competition increases in the gas marketplace, suppliers will need to reduce prices and improve communications with their customers. Regardless, some customers will continue to pay more than the spot market price to ensure a reliable supply of gas.

Bibliography

Ewing, Terzah, "Strong Dose of Winter Spikes Price of Natural Gas, Reversing Recent Losses," *The Wall Street Journal*, November 6, 1998, page C17.

Kennedy, John L., *Oil and Gas Pipeline Fundamentals 2nd Edition*. Tulsa: PennWell Publishing Company, 1993.

Morris, Kenneth M., and Siegel, Alan M., *The Wall Street Journal Guide to Understanding Money & Investing*. Lightbulb Press, Inc., 1993.

RKS Research & Consulting, National Survey of Business Customers, 1998.

Schlesinger, Ben, Benjamin Schesinger and Associates, Inc. Presentation at the Conference on Pipeline Industry Technology–20th Century Problems, 21st Century Solutions, Englewood, Colorado, May 13-14, 1997.

10

FUTURE SUPPLY AND DEMAND FOR NATURAL GAS

Introduction

When you pay your electric bill or fill up your tank with gasoline, you might think that energy is expensive. But in the context of the past few decades, energy prices are at relatively low levels. This means that people tend to use more energy rather than conserve it, because demand is strongly affected by price. Future supplies of natural gas and other limited energy resources will depend on how much energy we consume, how much we attempt to preserve for future generations, and how well we learn to develop and use our remaining resources efficiently.

In contrast to the 1970s, when people feared that energy supplies would soon run out, most experts believe the world contains enough gas, oil, and coal to support growth in populations and economies for many decades. Given this feeling of abundance, it will be difficult to restrain increases in energy consumption during the foreseeable future. Fortunately, the outlook for natural gas supply in the U.S. is quite optimistic, and the industry is poised to meet the challenges of satisfying America's growing appetite for gas.

Recent Trends

To forecast supply and demand for natural gas, analysts look first toward the past, to detect trends in gas production and consumption and then extrapolate them into the future.

Consumption

In 1996, natural gas accounted for 24% of America's total energy demand, and consumption reached about 21.9 trillion cubic feet (Tcf) (623 billion cubic meters, or 10^9 m³) (Table 10–1). This is nearly equal to the amount consumed in 1972, when gas consumption recorded its all-time high.

Table 10–1. Natural Gas Consumption by Sector

	1995		1996		
Sector	Tcf	10^9 m³	Tcf	10^9 m³	% of Total
Residential	4.9	138	5.3	149	23-24
Commercial	3.0	85	3.1	88	14
Industrial	9.5	270	9.8	278	44-45
Electricity Generation	3.4	96	3.0	85	16-14
Transportation	0.7	19	0.7	19	3
Total	**21.7**	**614**	**21.9**	**623**	—

Source: GRI 1998 Baseline Projection (numbers may not add due to rounding).

Residential demand for gas peaked during the 1970s, then steadily declined due to changes in building practices, improvements in equipment efficiency, and consumers' response to higher gas prices in the late 70s and early 80s. Since then, gas heating in single-family homes has rebounded, leading a resurgence of residential demand. Almost 70% of gas in this sector is used for heating. Similarly, heating accounts for 55% of gas consumed in the commercial sector.

During the mid-1980s, gas consumption in the industrial sector declined as customers switched to cheaper fuels, but this trend reversed as gas prices became more competitive. In 1996, the largest amount of gas in the industrial sector was used for process heat and steam.

From 1985 to 1996, gas consumption for electricity generation grew quite rapidly. Cogeneration (the production of power and useful heat at the same time) accounted for most of this growth (more than 80%).

Production

The bulk of natural gas in America is produced domestically (Table 10–2) in the Lower 48 States. As with gas consumption, production of gas peaked in 1972. Traditionally, Texas and Louisiana have produced the most natural gas, with much of it coming from offshore wells. In 1997, gas well completions jumped 24% over 1996, to 10,775 completions, and exceeded oil well completions for the fifth consecutive year.

Table 10–2. Current Natural Gas Supply

Source	1995 Tcf	1995 10^9 m^3	1996 Tcf	1996 10^9 m^3
U.S. Production	18.6	526	18.8	532
Imports	2.8	80	2.9	83
Supplements	0.5	15	0.1	3
Total	21.9	622	21.8	617

Source: GRI 1998 Baseline Projection (numbers may not add due to rounding).

Declining production from the Gulf states has been offset by increases in other areas, most notably the mountain states of New Mexico, Colorado, and Wyoming. The small amount of gas imported to America comes from Canada and Mexico, although some liquefied natural gas (LNG) is also imported from other countries. The U.S. exports small amounts of gas by pipeline to Canada and Mexico and as LNG from Alaska to Japan.

Current reserves

The U.S. contains an estimated 167 Tcf (4.7 trillion m^3) of natural gas reserves, accounting for 3.3% of the world's total. Evaluation of gas reserves requires calculating the amount of gas recoverable from every known reservoir. To qualify as "proved" reserves, the gas must be economically recoverable with current technology.

Proved gas reserves increased steadily throughout the 1960s and declined slowly during the 1970s as exploration and development activities by producers were curtailed. This trend was reversed when wellhead gas prices were deregulated, and since then the level of proved reserves has stabilized. Currently, annual production in the U.S. exceeds net addition to reserves; in other words, we are using more than we are finding. However, proved reserves will provide for more than eight years of production at current levels. Gas discoveries in 1996 totaled more than 12 Tcf (340 10^9 m³), which was 12% above 1995's level.

The offshore areas close to the U.S. coastline are believed to contain large amounts of the nation's recoverable oil and gas. Proved offshore reserves, which amounted to about 34.8 Tcf (986 10^9 m³) as of 1995, account for more than 20% of total U.S. reserves. In 1996, more than two-thirds of gas discoveries were in Texas and the offshore Gulf. The technology of drilling offshore in deep water has developed rapidly. Much success has also been realized in laying gathering lines to bring the gas to onshore processing plants and transmission lines. Nevertheless, offshore production continues to be very expensive.

In Alaska, major gas and oil reserves were not discovered until the late 1960s, when a giant oil field was found at Prudhoe Bay. This field's proved reserves of natural gas were estimated in 1995 at 9.5 Tcf (270 10^9 m³). However, most of the gas produced there is reinjected to maintain pressure in the field's oil wells. The gas industry is exploring the possibility of building a pipeline to carry the gas southward.

Future Supply and Demand

The outlook for natural gas supply in the U.S. is quite optimistic. In the long run (through 2020), gas production is expected to increase sharply, thanks to abundant reserves and improved technology for recovering offshore gas and developing unconventional resources. Between 2000 and 2005, natural gas will outstrip oil as the dominant revenue source for American producers – a pivotal change for the oil and gas industry. Increased gas production is expected to come mainly from onshore, nonassociated sources (those not in contact with oil underground). Offshore Gulf of Mexico production is also forecast to grow significantly, and Alaskan fields still contain a huge potential resource.

However, the gas industry will encounter challenges in meeting the anticipated growth in demand. U.S. natural gas consumption is expected to expand substantially for the next 15 to 25 years and will require a corresponding expansion of gas pipeline and storage capacity. In 1999 alone, U.S. demand for natural gas will jump by 5.2% over 1998's level. Through 2015, gas demand is projected to rise by 2% each year, reaching a total of about 31 Tcf (880 10^9 m³) (see Table 10-3).

During this period, the natural gas share of total U.S. energy consumption will increase to 28% from the current level of 24%. This demand growth translates to a total increase of 9 Tcf (250 10^9 m³), or an annual growth of about 0.5 Tcf (10 10^9 m³). The last time the gas industry faced a similar challenge was from the mid-1950s to early 70s, when growth increased 13 Tcf (350 10⁹ m³), amounting to 0.8 Tcf (20 10⁹ m³) per year.

Table 10–3 Natural Gas Demand by Sector

Sector	2000 Tcf (10^9 m³)	2005 Tcf (10^9 m³)	2010 Tcf (10^9 m³)	2015 Tcf (10^9 m³)
Residential	5.2 (146)	5.4 (152)	5.5 (157)	5.8 (165)
Commercial	3.3 (94)	3.5 (99)	3.7 (105)	4.0 (113)
Industrial	10.4 (295)	11.3 (320)	12.0 (339)	12.6 (358)
Electricity Generation*	3.6 (102)	4.6 (129)	5.4 (154)	6.9 (196)
Transportation				
Pipelines	0.8 (22)	0.9 (25)	1.0 (28)	1.1 (30)
Vehicles	—	0.1 (3)	0.4 (11)	0.6 (17)
Total	23.2 (658)	25.7 (727)	28.0 (793)	31.0 (879)

* Does not include gas used in commercial and industrial cogeneration systems.

Source: GRI 1998 Baseline Projection (numbers may not add due to rounding).

Much of this growth in gas demand will be fueled by environmental concerns and competition in the energy marketplace. To slow down the process of global warming, the U.S. and other industrial nations worldwide are committed to reducing emissions of carbon (primarily in the form of carbon dioxide). In October 1997, President Clinton said that America must pursue a policy of "fuel conversion" from coal to natural gas for electric power generation.

Many of the actions that could be taken to control these emissions are already happening, thanks to more intense competition among energy suppliers. When fuel is converted to useful energy, higher efficiency processes yield not only lower emissions, but also lower costs. For example, in the electric generation sector, competition is driving efficiency improvements, as existing facilities are being modified or replaced and combined cycle plants are being constructed.

Demand by sector

In the residential sector, the use of gas for heating more or less determines demand, since heating dominates residential consumption. Gas heat will face more intense competition in the future, resulting in a tapering off of demand. However, growth in non-heating uses for gas (cooking, water heating, and clothes drying) is expected to offset this slight decline, resulting in a gradual increase in residential demand through 2015 (Table 10–3). Applications for gas such as hearth products and air conditioning will also contribute to growth. Over the same period, the number of residential gas customers will increase to almost 70 million.

In the commercial sector, the bright spot for gas is cooling (air conditioning and refrigeration). Gas cooling and other non-heating uses for gas, such as cooking and water heating, will account for almost 90% of the projected growth in commercial gas demand through 2015. Because of technology development, gas is expected to capture a larger share of the commercial cooling market and could account for about 17% of the total in 2015. The gas share of cooling equipment sales is expected to be just over 20% annually from 1996 to 2005.

Strong growth is also projected for industrial gas demand. The use of gas to produce process steam accounts for much of this growth in demand, and steam applications will overtake gas demand for process heat. However, gas is projected to continue to dominate industrial process heat applications, holding on to roughly a 55% share. In addition to these uses, gas-fueled cogeneration of heat and power by industrial customers (in boilers and other systems) will also spur healthy demand growth, mostly in the earlier years (through 2005).

The fastest growth in U.S. gas demand will result from new gas-fueled power plants for electricity generation. In particular, combined cycle facilities that use more efficient gas turbines will help lower the cost of gas-generated electricity to levels competitive with coal-fired plants. Overall, gas could account for about 20% of the energy used to produce electricity by 2015, but for the foreseeable future, coal will dominate energy demand for power generation.

Natural gas is frequently the fuel of choice by independent (non-utility) power producers, because they usually build relatively small plants and want to avoid problems with air quality regulations and permitting. Independent power generators are anticipated to rapidly increase their share of total electricity production through 2015. When combined with growth in gas demand for cogeneration, the non-utility share could increase to 27% of total U.S. power production.

Gas demand in the transportation sector could increase mainly due to greater numbers of natural gas vehicles on the road, while gas use at pipeline compressor stations grows only slightly. Most of the increase in vehicle demand for gas is accounted for by heavy-duty trucks.

Supply and price

Gas prices in real terms (accounting for inflation) are only about half of what they were in the mid-1980s, yet North American gas production is at record levels, and producers are poised to break those records. Despite flat real prices through 2015, the U.S. gas supply is projected to increase by about 40%, from just under 22 Tcf (about 620 10^9 m³) in 1995-96 to 31 Tcf (879 10^9 m³) (Table 10–4). The gas industry will continue to accelerate pipeline construction in North America to transport larger volumes of gas.

Table 10–4. Future Natural Gas Supply

Source	2000 Tcf (10^9 m³)	2005 Tcf (10^9 m³)	2010 Tcf (10^9 m³)	2015 Tcf (10^9 m³)
U.S. Production	19.8 (559)	21.1 (598)	23.5 (667)	26.4 (746)
Imports	3.3 (94)	4.5 (127)	4.5 (127)	4.5 (127)
Supplements	0.2 (5)	0.2 (5)	0.2 (5)	0.2 (5)
Total	23.3 (658)	25.8 (730)	28.3 (802)	31.0 (879)

Source: GRI 1998 Baseline Projection (numbers might not add due to rounding).

Production from the Lower 48 States will continue to dominate U.S. gas supply and by 2010 will exceed its 1972 peak. Afterward, growth in production will come from less conventional gas resources, principally low permeability reservoirs. Production from additional offshore gas resources and deep onshore gas will depend on development of new technology. Gas drilling activity is expected to increase rapidly, but not until after 2010. Even then, the number of gas wells is not projected to approach historically high levels. In 2015, about 17,000 gas wells are anticipated to be drilled, compared to 1981's peak of 20,000.

Natural gas prices are expected to remain stable through 2015 in real terms (Table 10–5), given flat oil prices and declining prices for coal and electricity. Compared to actual prices in 1995-96 of about $1.50-$2.00 per million Btu (billion Joules), gas should remain a bargain in 2015 at an average price of $1.95. These are "acquisition" prices, which do not include the cost of transportation (transmission and distribution).

The gas price that customers pay should actually fall over the next 20 years. Thanks to more efficient, less expensive transportation of gas, the average "burner-tip" price is projected to decline by about 15%, from $4.06/million Btu (billion Joules) in 1996 to just $3.41 in 2015.

Potential Gas Resources

In addition to America's proved reserves, more natural gas will be found in the U.S. in new fields and in existing areas, which are routinely extended as geologists define additional productive zones. Government agencies and various gas industry organizations prepare estimates of these "potential" natural gas resources. The Potential Gas Committee has established criteria for estimating and classifying gas resources, as follows.

Table 10–5. Average U.S. Gas Prices ($/million Btu or $/billion Joules, in 1996 Dollars)

	Acquisition	Burner-Tip
1995	1.55	3.58
1996	2.06	4.06
2000	1.94	3.83
2005	1.87	3.63
2010	1.91	3.52
2015	1.95	3.41

Source: GRI 1998 Baseline Projection (numbers might not add due to rounding).

Probable resources

Probable resources are associated with known fields and are the most likely to be proven. A relatively large amount of geologic and engineering information is available to help estimate this resource. Probable resources bridge the boundary between discovered and undiscovered gas.

The discovered portion includes gas supply from future extensions of existing pools in known productive reservoirs. The pools containing this gas have been discovered, but their extent has not been completely delineated by development drilling. Therefore, the existence and quantity of gas in the undrilled portion of the pool are as yet unconfirmed.

Possible and speculative resources

Possible resources are a less reliable supply because they are postulated to exist outside known fields, but they are associated with a productive formation.

Their occurrence is indicated by a projection of plays or trends into a less well-explored area. Possible resources are expected to be found from new field discoveries within these trends or plays. The types of traps and/or structural settings might be the same or might differ in some respect.

Speculative resources, the most nebulous category, are expected to be found in formations or provinces that have not yet proved to be productive. Geologic analogs are developed to ensure reasonable evaluation of these unknown quantities. The resources are anticipated from new pool or new field discoveries.

Total potential

Potential resources are estimated for onshore and offshore locations separately for the Lower 48 States and Alaska (Table 10-6). Potential resource esti-

Table 10–6. Estimated Potential Natural Gas Resources, Tcf (10^{12} m³)

Traditional	Probable	Possible	Speculative	Total
Lower 48 States				
Onshore	125.2	167.4	138.5	431.1
	(3.5)	(4.7)	(3.9)	(12.3)
Offshore	17.3	57.8	73.7	148.8
	(0.5)	(1.6)	(2.1)	(4.2)
Subtotal	142.5	225.2	212.2	579.9
	(4.0)	(6.4)	(6.0)	(16.4)
Alaska				
Onshore	31.3	16.4	27.7	75.4
	(0.9)	(0.5)	(0.8)	(2.1)
Offshore	2.4	12.7	53.0	68.1
	(0.07)	(0.4)	(1.5)	(1.9)
Subtotal	33.7	29.1	80.7	143.5
	(1.0)	(0.8)	(2.3)	(4.1)
Total Traditional	**176.1**	**254.3**	**292.9**	**723.3**
	(5.0)	(7.2)	(8.3)	(20.5)
Coalbed Methane	12.8	38.2	83.2	134.2
	(0.4)	(1.1)	(2.4)	(3.8)
Total U.S.	**189.0**	**292.5**	**376.1**	**857.5**
	(5.4)	(8.3)	(10.6)	(24.3)

Source: Potential Gas Committee, 1990.

mates also include volumes of methane that may be recoverable from coal seams, an increasingly important source of natural gas.

Although additional resources might become available with advanced technology and changes in economic conditions, these estimates do not include natural gas that could be recovered from unconventional sources such as hydrates or very low-permeability formations. The estimates also exclude very deep gas (in excess of 30,000 feet, or 10,000 meters) and offshore reserves in water deeper than 3,000 feet (1,000 meters), except for the Gulf of Mexico, where gas is considered accessible in deeper water.

Bibliography

American Gas Association, *Gas Facts 1998*, www.aga.com.

Gas Research Institute, *1998 Policy Implications of the GRI Baseline Projection of U.S. Energy Supply and Demand*, www.gri.org.

United States Energy Information Administration, April 1998, www.eia.doe.gov/cabs/usa-html.

GLOSSARY

Abiogenic Gas - Natural gas that might have been formed by nonbiological processes, without organic matter. This theory of gas origin is not widely accepted.

Absorption Chiller - Gas-powered equipment used to air condition commercial buildings through an absorption process. Water is used as the refrigerant instead of chemicals that contribute to global warming.

Acid Gas - Natural gas containing carbon dioxide or hydrogen sulfide. These impurities can form acids that corrode metal pipe. Acid gas is conditioned by a sweetening process.

Acoustic Velocity Log - *See Sonic Velocity Log.*

Acquisition Price - The price of natural gas not including the cost of transmission and distribution.

Allocations - The means by which pipeline companies prioritize their customers according to system capacity, which is allocated first to firm customers and then to interruptible customers.

Alternative Fuel Vehicle - Vehicle that can operate on an alternative fuel, usually something other than gasoline or diesel fuel. Alternative fuels include natural gas, ethanol, and methanol.

Amplitude Variation with Offset (AVO) - A technology used to enhance analysis of bright spots in seismic data.

Angular Unconformity - A type of stratigraphic trap that results from a sequence of geological processes and can capture huge amounts of gas if overlain by an impermeable cap.

Anticline - A type of structural trap where layers of rock have been gently folded upward to form an arch.

Aquifer - Water-bearing formations used to store natural gas underground.

Associated Gas - Natural gas found in contact with crude oil below ground, either in the impermeable cap rock or actually dissolved in the oil. Associated gas contains many other hydrocarbons besides methane.

Avoided Cost - The price at which an electric utility must buy power back from its own customers, usually industrial plants. The avoided cost was determined by federal legislation to be the same as the electric utility's cost of generating power.

Back Offices - The accountants and other administrators who independently verify a gas contract or agreement between two parties.

Base Load - The bulk of natural gas demand, representing normal levels of gas consumption.

Basis Trading - The method used by gas traders and brokers on the financial market, based on the difference in the price of natural gas at different locations on the pipeline. This value, called the market basis differential, differs depending on local supply/demand situations.

Batch - Raw glass material used in manufacturing glass products.

Bi-fuel Vehicle - Vehicle that can operate on either natural gas or conventional fuel (gasoline or diesel). Also called dual-fuel vehicle.

Biogenic Gas - Almost pure methane that is formed by biological processes (bacterial action) at relatively low temperatures and depths. Also called microbial gas, swamp gas, or marsh gas.

Bit - *See Drill Bit.*

Blowout - An uncontrolled flow of gas from a well. Blowout preventers are at the top of the well to close it off when a kick is detected.

Boiler - Commercial or industrial equipment used to produce steam or hot water.

Booster Water Heater - Natural gas water heater designed to raise the temperature of water for commercial dishwashing to the level required by government regulations.

Bright Spot - An area of intense reflection on a seismic profile which usually indicates the location of natural gas reservoirs and caps of gas lying over oil fields.

Build Angle - The curved portion of a directional or horizontal well following the kickoff point. Also called the *dogleg*.

Burner-tip Price - The price that the end user (customer) pays for natural gas at the burner tip, including the costs of transmission and distribution.

Butane - A component of natural gas with the chemical formula C^4H^{10}. Propane and butane are usually extracted from natural gas and sold separately.

By-pass - The purchase of gas directly from producers and marketers, rather than through a pipeline company. Federal regulations allow large gas customers and distribution companies to by-pass pipelines.

Cable-tool Drilling - A method of drilling a well by pounding the earth with a chisel-like drill bit.

Caliper Log - A wireline well log that measures wellbore diameter, indicating the type of rock being drilled.

Cap Rock - Impermeable rock layer that prevents gas from seeping upward out of a trap.

Carbon Dioxide - A by-product of natural gas combustion with the chemical formula CO^2. Carbon dioxide also occurs as an impurity in some natural gas fields.

Carbonate - A type of reservoir rock, usually calcium carbonate (limestone).

Casing - Thin, seamless steel pipe that is used to line a well. Casing is installed in pieces as the well is drilled and then at the bottom once the well proves likely to produce gas.

Cathodic Protection Systems - Electrical devices installed on metal piping underground to protect against corrosion by inhibiting electrochemical reactions between the pipe and the soil.

Cement - Wet slurry that is pumped between the casing and the sides of the well to create a foundation.

Centrifugal Compressor - Type of compressor used by pipelines to maintain gas pressure during transportation.

Checkers - Bricks used inside glass melting furnaces to capture exhaust heat.

Chiller - Equipment used to air condition commercial buildings.

Chlorofluorocarbons (CFCs) - Chemicals containing chlorine, such as freon and other refrigerants. CFCs are considered to be the main culprit in depletion of the earth's protective ozone layer.

Christmas Tree - A series of valves and fittings installed on the wellhead to control gas flow. Also called a *production tree*.

City Gate Station - The point where natural gas is received by the distribution company from pipelines or other sources of supply. Also called *town border* or *tap stations*.

Clamshell - A double-sided broiler/griddle used for commercial cooking.

Claus process - A method of recovering sulfur from natural gas.

Coal - A solid fossil fuel composed of hydrocarbons. Coal is formed when woody material is transformed by temperature and time.

Coal gas - Combustible fuel made by heating coal. Also called *manufactured gas or town gas*.

Coalbed Gas - *See Coal Seam Gas.*

Coal Seam Gas - Natural gas (virtually pure methane) created when woody material was changed into coal. The gas is adsorbed to the surface of the coal along natural fractures. Also called *coalbed gas*.

Cofiring - A process using natural gas as a supplemental fuel in power plant boilers to reduce emissions of sulfur dioxide and particulates.

Cogeneration - The production of electricity and heat (hot water or steam) at the same time. Typically, cogeneration processes are powered by a natural gas engine or a boiler-turbine combination. Also called *combined heat and power*.

Coke - A solid, porous by-product of gas manufacturing that can be used for domestic heating. Coke is also used in iron and steel production.

Combination Trap - Petroleum traps that have both structural and stratigraphic elements.

Combined Cycle - Type of power plant that uses a mix of gas and steam turbines.

Commodity - Raw material traded on financial exchanges worldwide, such as natural gas, oil, grain, and metals.

Completion - The process of completing a well by lining the bottom of the bare hole with steel casing, setting its foundation underground with cement or other materials, and punching holes through the casing and cement to allow gas to flow into the well.

Compression Ratio - The ratio of outlet to inlet pressure on a pipeline compressor.

Compressor - Device used to increase the pressure of natural gas during transmission. Also, device used in chillers to produce cold water for air conditioning.

Condensate - Heavier, liquid hydrocarbons that can exist as gases underground but which re-liquefy, or condense, when the gas is produced. *Also called natural gasoline*.

Conditioning - Processes that remove water and impurities from natural gas. Gas conditioning processes include sweetening to remove carbon dioxide and hydrogen sulfide and glycol dehydration to remove water.

Confirmations - Contractual operations that gas traders perform to manage the risk of their deals. A confirmation is an independent verification of a contract by the accountants and other administrators from both parties.

Controlled Atmosphere Furnace - A type of heat treating furnace that uses a controlled atmosphere inside to protect the metal product from oxidation or promote a particular chemical reaction.

Core - A sample section extracted from an underground rock formation and used to develop information on the rock's ability to produce gas.

Corrosion - Damage caused to metal piping by acid or water inside the pipe or by electrical differences between the pipe and the surrounding soil.

Cross-well Seismic - A technology that uses a seismic energy source in one well and signal receivers in one or more nearby wells. Cross-well images have greater resolution than surface seismic data.

Crude Oil - Oil as found underground before refining. Crude oil occurs as a liquid and is composed of more than 100 types of hydrocarbon molecules.

Cryogenic - Very low temperatures used to store liquefied natural gas (LNG).

Cullet - Scrap glass material used in manufacturing glass products.

Curtailment - The practice of cutting off certain customers when gas supply cannot meet demand, which occurred during the 1960s-1970s. Usually, the first customers to be curtailed were power plants and large industrial facilities.

Cushion gas - A permanent volume of gas kept in a storage reservoir to maintain sufficient pressure for gas production.

Cuttings - Chips of rock collected while drilling gas wells and used to develop information on the rock's ability to produce gas.

Dedicated Vehicle - Vehicle that operates only on natural gas or other alternative fuel and cannot use gasoline or diesel fuel.

Deformation - Geological changes where intense pressure causes rocks to rise, sink, or shift from side to side. Rocks are also deformed by weathering and erosional processes that transport and deposit sediments.

Demand Charge - Extra cost for electricity during peak demand periods, usually daytime and summer.

Depleted Reservoir - Underground reservoir that has been emptied of commercial amounts of gas or oil. These reservoirs are often used for natural gas storage.

Derrick - A four-legged tower that sits on the floor of a drilling rig.

Desiccant - Drying agent such as silica gel used to remove moisture from outdoor air before it enters a building's heating and cooling system.

Dip Log - A wireline well log that determines the orientation of rock layers.

Dip-slip Fault - A fault where the rock sections have shifted up and down, as opposed to sideways (strike-slip).

Directional Drilling - Deviating from straight vertical drilling to reach a specific target.

Discovery Well - The first exploratory well that discovers gas in a new field.

Distribution Main - Piping that carries natural gas underground to individual customer service lines.

Distribution System - Piping network that carries natural gas to its customers from various sources of gas supply.

Dogleg - The curved portion of a directional or horizontal well following the kickoff point. Also called the *build angle*.

Dome - An uplift structure similar to an anticline, where gas is trapped at a high point in the reservoir rock.

Drill Bit - The tool at the bottom of a well that excavates the rock. Most drill bits are shaped like three cones melded together, with teeth along the cone edges.

Drilling Barge - A type of offshore rig used mainly in shallow, protected waters.

Drilling Mud - Fluid pumped down a well to clean and lubricate the rotating drill bit. The mud flows back to the surface through the annular space between the rotating drillstring and the borehole walls, carrying cuttings from underground formations.

Drilling Time Logs - Measurements recording the drill bit's rate of penetration.

Drillpipe - Sections of steel piping that are threaded together as they go down into a well.

Drillship - A ship designed to drill offshore through a hole in the hull.

Drillstring - Part of the rig's rotating equipment, including the turning drillpipe and bit.

Dry Gas - Pure methane that forms no liquid condensate either in the reservoir or aboveground.

Engine Driven Chiller - Gas-powered equipment used to air condition commercial buildings. The engine replaces the electric motor to supply power to the compressor.

Ethane - A component of natural gas with the chemical formula C^2H^6.

Expander Plant - Equipment used to separate liquid condensate from dry gas.

Exploratory Well - A well drilled to discover a new gas field. Also called a *wildcat well*.

Fault - Structural traps where the rocks have been fractured and large sections have slipped past one another.

Feeder Main - Piping that transports gas from the supply main or pressure regulator to the distribution mains.

Feedstock - Raw material such as natural gas used to manufacture chemicals made from petroleum.

Fish - Pieces of metal that break off inside a well or tools that fall down into a well accidentally. Also called *junk*. Drilling must be suspended until fish can be retrieved with special tools leased from a service company.

Fixed Leg Platform - An offshore drilling platform with legs held in place by piles driven into the sea bottom.

Flat Spot - A reflection off of a gas-oil or gas-water contact on a seismic profile.

Fluid - Liquid or gas; specifically, oil or gaseous natural gas.

Fold - Structural traps where layers of rock have been gently folded, including anticlines, domes, and synclines.

Force Majeure - Pipeline contract options that allowed companies to prove in court that extremely unusual market conditions are similar to a natural disaster. Also called *market out options*.

Forced Draft Fan - Fan that pushes combustion air into residential warm-air furnaces.

Formation - The basic rock layer used for geologic mapping. Formations have a definite top and bottom and are classified with a two-part name indicating geographic location and dominant rock type.

Fossil Fuel - Fuels that were created by the decomposition of organic matter. Fossil fuels include oil and natural gas (petroleum) and coal. Fossil fuels are considered non-renewable, in contrast to other energy resources that can continue to be produced, such as solar and wind energy.

Fracturing - Increasing the permeability of a tight formation by injecting hydraulic fluids to shatter the rock and enlarge its flow passages around the wellbore. Also called *stimulation*.

Gamma Ray Log - A wireline well log that measures radioactivity, which indicates the type of rock in the well.

Gamma-gamma Log - A wireline log that measures rock porosity. Also called a formation *density log*.

Gas Conditioning - *See Conditioning*.

Gas Cooling - Air conditioning and refrigeration powered by natural gas. Gas cooling equipment includes absorption chillers, engine driven chillers, and engine driven refrigeration systems.

Gas Marketer - Person acting as middleman between a gas customer and the source of gas supply (producer, pipeline company, etc.). The marketer actually takes title to the gas for at least a brief period, in contrast to traders or brokers who simply make the arrangements.

Gas Trader - Person acting as middleman between a gas customer and the source of gas supply (producer, pipeline company, etc.). The trader simply makes arrangements between the two parties without actually taking title to the gas. Also called day *trader* or *broker*.

Gas Well - *See Well*.

Gathering System - A network of flow lines collecting gas from several wells and transporting it to a central location for processing.

Geochemical Methods - Exploration techniques based on analyzing the chemical and bacterial composition of soil and water on the surface, above or near underground gas reservoirs. These methods can reveal invisible seeps of petroleum that often occur in a halo pattern.

Geological Methods - Exploration techniques that involve drawing maps of surface and subsurface structures and taking samples of rock formations.

Geology Related Imaging Program (GRIP) - A method of increasing the resolution of a seismic image by incorporating geological information directly into the seismic survey design.

Geophones - Vibration sensors on the surface that detect returning echoes from seismic waves and translate them into electrical voltage. Also called *jugs*.

Geophysical Methods - Exploration techniques that measure the physical characteristics of underground rock formations, such as seismic reflection, gravity, and magnetic force. Geophysical techniques enable geologists to determine the depth, thickness, and structure of subsurface rock layers and evaluate whether they are capable of trapping natural gas.

Glycol - A liquid desiccant used to remove water from natural gas.

Gravel Pack - Coarse sand used to complete wells in unconsolidated sands.

Gravity Meter - A geophysical exploration tool used to evaluate underground rock density.

Greenhouse Gas - Gases that might contribute to global warming, such as carbon dioxide and methane.

Heat Treating - The controlled heating and cooling of a metal or alloy, usually to give the product certain desirable properties.

Helium - A trace constituent of natural gas found mainly in just a single U.S. gas field.

Holiday - A defect in the coating on metal pipe that can lead to corrosion.

Horizontal Drilling - Changing direction underground until the wellbore is penetrating the formation laterally. Horizontal wells can increase production from thin formations and low-permeability reservoirs.

Hybrid System - Commercial air conditioning system that contains a mix of electric and gas-powered equipment.

Hydraulic Fracturing - Increasing the permeability of a tight formation by injecting hydraulic fluids to shatter the rock and enlarge its flow passages around the wellbore. Also called *stimulation.*

Hydrocarbon - Chemical compounds containing primarily carbon (C) and hydrogen (H). Crude oil and natural gas are composed of hydrocarbon molecules.

Hydrogen Sulfide - A very poisonous, foul-smelling gas that occurs as an impurity in some natural gas fields. Hydrogen sulfide can form sulfuric acid, which corrodes metal pipe, and is removed from natural gas by conditioning processes.

Hydrophones - Vibration detectors used in offshore seismic operations.

Independent Power Producer - Unregulated, non-utility companies that build and operate power plants, often using cogeneration technology to produce useful heat as well as electricity.

Induced Draft Fan - Fan that pulls combustion air into residential warm-air furnaces.

In-fill Well - Additional wells drilled to increase gas production rate from an existing field.

Infrared Burner - A type of natural gas burner, either metallic or ceramic, that uses infrared heat for industrial drying processes.

Injection Well - Well used to inject natural gas into a storage reservoir.

Interruptible Customer - An energy user who agrees to allow a temporary supply cut-off (interruption) in exchange for a lower energy rate. Usually, the customer's supply of natural gas or electricity is cut off only during times of extreme peak demand.

Isopach Map - A type of geologic map showing the thickness of rock layers.

Jack-up Rig - A type of offshore drilling rig with legs that can be raised and lowered.

Joints - Sections of large-diameter transmission pipe. Also, fused connections that join pieces of plastic distribution piping.

Kelly - A piece of square pipe that is gripped and turned by a drill rig's rotating system.

Kick - An encounter with unexpectedly high pressures underground, which can let gas or water flow into the well, dilute the drilling mud, and reduce its pressure.

Kickoff - The spot in a directional or horizontal well where the borehole begins to deviate from the vertical.

Kill Mud - Heavier drilling mud pumped down to circulate a kick out of a well.

Leaching - The process of creating natural gas storage caverns by pumping water underground to dissolve the rock.

Leak Detector - Device used to inspect or survey natural gas piping for leakage.

Limestone - A common sedimentary rock composed of calcium carbonate. It can range from fine to coarse grained and can be a reservoir rock.

Linepack - Unsold gas built up in a pipeline.

Liquefied Natural Gas (LNG) - Natural gas that has been chilled to very cold temperatures, which liquefies it for storage or for use as a fuel in heavy-duty vehicles.

Liquefied Petroleum Gas (LPG) - Consisting mainly of propane, LPG is a common substitute for natural gas in rural areas not served by pipelines.

Liquid Redox Process - A method of recovering sulfur from natural gas based on reduction/oxidation.

Liquidity - The flexibility with which an asset, such as a commodity or share of stock, can be turned into cash.

Lithification - The process of compaction of loose sediments into rock.

Lithofacies Map - A type of geologic map showing the variations (facies) in a single rock layer.

Lithographic Log - Sampling of the well cuttings flushed up with the drilling mud.

Logs - Measurements and other information recorded while a well is being drilled, whether for gas or other resources.

Magnetometer - A geophysical exploration tool used to detect variations in the elevation of basement rock, estimate thickness, and locate faults.

Manufactured Gas - Combustible fuel made by heating coal. Also called coal gas or town gas. The early gas industry in America and Europe was based on manufactured gas, not natural gas.

Market Basis Differential - The difference in the price of natural gas at one location on the pipeline and at another location. A molecule of gas in different locations is worth different values, depending on local supply/demand situations.

Market Out Provisions - Pipeline contract options that allowed companies to prove in court that extremely unusual market conditions are similar to a natural disaster. Also called *force majeure*.

Marketer - *See Gas marketer.*

Marsh Gas - Almost pure methane that is formed by biological processes (bacterial action) at relatively low temperatures and depths. Also called microbial gas, swamp gas, or biogenic gas.

Meter - Device used to measure the volume of gas passing through a certain point in a pipeline or distribution system.

Methane - The simplest hydrocarbon molecule, with the chemical formula CH_4, and the main component of natural gas. Methane is a colorless, odorless gas that burns readily with a pale, slightly luminous flame. In addition to burning the gas as a fuel, methane is a key raw material for making solvents and other organic chemicals.

Microbial Gas - Almost pure methane that is formed by biological processes (bacterial action) at relatively low temperatures and depths. Also called *biogenic gas*.

Migration - The vertical and horizontal flow of gas from its source rock. Because gas is light in density, it flows upward along fractures and faults, or it can flow horizontally and then upward through permeable rock layers.

Mineral Rights Owner - Individual or government controlling the right to explore, drill, and produce gas or other resources.

Modeling - Mathematical computer programming that help geologists imagine underground structures and conditions, rather than measure their properties.

Modulating Burner - Natural gas burners that adjust their heat output gradually instead of cycling on and off.

Mud Logs - Analysis of the chemistry of the drilling mud and cuttings to detect traces of natural gas.

Natural Gas Liquids (NGL) - Heavier hydrocarbons including condensate, butane, propane, and ethane.

Natural Gasoline - Heavier, liquid hydrocarbons that can exist as gases underground but which re-liquefy, or condense, when the gas is produced. Also called *condensate*.

Neutron Log - A wireline log that measures rock porosity.

Nomination - A contractual operation in which the gas customer tells the pipeline company 24 hours in advance exactly how much gas should be delivered and where to deliver it. Daily nominations help keep the pipeline system balanced.

Nonassociated Gas - Natural gas that does not contact crude oil in the trap. Nonassociated gas wells produce almost pure methane.

Odorization - The process of adding an artificial odor to natural gas. Odorization is required for safety, so that people can readily detect even minute gas leaks.

Off Peak - Times when demand for energy (gas or electricity) is at normal or low levels, as opposed to peak demand periods, when maximum supplies are required.

Oil - A liquid fossil fuel often found along with natural gas underground.

Onsite Power - Generation of electricity at a manufacturing plant. Also called *customer-sited* or *distributed power.*

Opacity - Clouds of particulate matter released by utility power plants.

Open Access - The requirement for pipelines to transport or store gas for customers who purchase supplies directly from producers or marketers.

Organic Matter - Compounds containing carbon (C), which can be decomposed over time to create natural gas and crude oil (hydrocarbons).

Orifice Meter - Common device used to measure the volume of gas passing through a certain point in a pipeline or distribution system.

Outcrop - Spots where rock formations are exposed aboveground.

Overthrust belt - A zone of thrust faults such as the Rocky Mountain range.

Oxygen-enriched Air Staging - A process used in glass melting to reduce emissions of nitrogen oxides.

Paleocave System - Carbonate rock reservoirs with very complex geology, formed from ancient caves connected by fractures.

Peak Demand - Natural gas demand in excess of normal levels. Also called *peak load.* Peak consumption of gas occurs for only short periods during the year, mainly on cold winter days. In contrast, peak demand for electricity occurs mainly on hot summer days.

Permeability - A measurement of how easy it is for fluids to flow through a rock.

Perforations - Holes shot through a well's casing and cement to allow gas to flow into the well.

Petroleum - Crude oil and natural gas.

Pig - A device used to clean and inspect the inside of gas transmission pipelines. Smart pigs use on-board computers for more accurate diagnosis of problem areas.

Pill - A mass of chemical additive injected into drilling mud to correct downhole problems.

Pipeline-quality Gas - Natural gas that has been treated to meet the specifications of a pipeline purchase contract.

Play - An area shown to contain commercial quantities of natural gas, with a proven combination of reservoir rock, trap, and cap rock or other seal.

Polyethylene - Plastic material used for natural gas pipe, including distribution mains and service lines.

Pores - Minute open spaces in rock that can contain fluids.

Porosity - A measurement of the capacity of a reservoir rock to hold fluids in its pores.

Potential Resources - The amount of gas that might be found in addition to proved reserves as geologists discover new fields and define new productive zones in existing fields. Potential resources include probable, possible, and speculative, according to their likelihood of being found and proved.

Preheating - Heating combustion air before fuel is burned. Raw materials can also be preheated. Preheating is often used to improve the efficiency of industrial processes.

Propane - A component of natural gas with the chemical formula C^3H^8. Propane and butane are usually extracted from natural gas and sold separately. Propane is the main component of liquefied petroleum gas (LPG).

Pressure Regulator - Device used to reduce the pressure of natural gas at various points throughout the distribution system.

Prospect - The exact location where geological and economic conditions are favorable for drilling an exploratory well.

Proved Gas Reserves - The amount of natural gas that is economically recoverable using current technology.

Purchase Contract - An agreement for a customer to buy gas from a producer, pipeline company, or gas marketer.

Purchased Gas Adjustment Clause - A mechanism for distribution companies to recover their cost of gas more quickly than through a conventional rate application. These clauses were initiated by state regulatory commissions during the 1970s when gas prices escalated rapidly.

Reburning - A process using natural gas as a supplemental fuel in power plant boilers to reduce emissions of nitrogen oxides.

Reciprocating Compressor - Type of compressor used by pipelines to maintain gas pressure during transportation.

Regenerative Burner - A type of natural gas burner that comes in pairs that fire in an alternating on/off sequence. The burner that is off recovers heat from the burner that is on and uses it to preheat combustion air, which reduces energy consumption.

Regulation - A rule or order having the force of law issued by an executive authority of a government. Generally, regulations are issued to implement or enforce laws or legislation enacted by the government.

Reservoir - An underground deposit of natural gas and/or crude oil. The petroleum is contained in the pores of a reservoir rock. Fluids (gas or oil) cannot flow from one reservoir rock to another.

Reservoir Pressure - The pressure on fluids in the pores of rock at a specific depth. Normal reservoir pressure is due to the weight of the overlying fluids or formations.

Reservoir Rock - A rock that has porosity and permeability, which give it the ability to hold and transmit fluids (gas or oil).

Resistivity Log - Measurement of the ability to conduct an electrical current. Rocks and their fluids (gas or oil) exhibit a characteristic resistivity, which can indicate the type of rock in underground formations.

Rig - The equipment used to drill a well, usually owned and operated by contractors.

Rock - An aggregate of mineral grains and crystals.

Rooftop Unit - A natural gas heating system installed on a building roof. Also called *unitary systems* or *packaged units*, as opposed to large, central heating systems.

Rotary Drilling - A method of drilling in which the rig spins a long length of steel pipe with a bit mounted on the end. Rotation of the bit digs into the earth and creates the wellbore.

Royalty - A percentage of the revenue from gas production that is paid to the mineral rights owner or others, free and clear of production costs.

Salt Dome - A type of combination trap where a large plug of salt has flowed upward through the overlying sedimentary rock.

Sandstone - A common sedimentary rock composed primarily of sand grains. Sandstone can be a reservoir rock.

Scrubber - Electrostatic precipitator used to reduce emissions from utility power plants.

Sediment - Loose particles of earth, mud, or salts. Sediments are deposited out of water, air, or ice.

Sedimentary Rock - A rock composed of sediments deposited on the surface of the ground or at the bottom of an ocean or other body of water.

Seismic Reflection - A geophysical exploration method using sound energy to characterize underground formations. The waves of sound or vibration are generated on the surface and reflected back upward by underground rock layers.

Semisubmersible - An offshore drilling platform that floats on submerged pontoons and is anchored at the drillsite.

Service Lines - Small diameter plastic piping that delivers natural gas from the main to the customer's meter.

Shale - A very common sedimentary rock, often rich in organic matter. Black shales are source rocks for petroleum.

Sonde - A cylinder filled with instruments that is run down into a well on a wireline to record a well log. It senses the electrical, radioactive, and sonic properties of the rocks and their fluids, as well as the diameter of the wellbore.

Sonic Amplitude Log - A wireline well log that measures the attenuation of sound through rocks to detect fractures.

Sonic Velocity Log - A wireline well log that measures the speed at which sound travels through the rock, which indicates porosity.

Sour Gas - Natural gas containing hydrogen sulfide, a very poisonous, foul-smelling gas that can form sulfuric acid and corrode metal pipe. Hydrogen sulfide is removed from natural gas by conditioning processes.

Source Rock - A sedimentary rock rich in organic matter that has been transformed by geological processes into natural gas and/or crude oil.

Spot Market - The short-term market for natural gas, created by deregulation of wellhead prices. The spot market allowed more competition, resulting in lower gas prices.

Spud - To break ground and begin digging a well.

Step-out Well - A well drilled to the side of a discovery well to determine the extent of a new field.

Stimulation - Increasing the permeability of a tight formation by enlarging the flow passages in the rock around the wellbore. Also called *fracturing*.

Stratigraphic Column - A diagram showing the vertical succession of rock layers, with the youngest at the top and the oldest at the bottom.

Stratigraphic Trap - A petroleum trap formed when the rock's permeability or porosity changes, preventing gas from migrating out of the rock. The gas is captured within the strata, or layers, of rock. Stratigraphic traps are generally harder to discover than structural traps.

Strike-and-dip - A map symbol showing the third dimension, that is the horizontal and vertical orientation of the rock layers.

Strike-slip Fault - A fault where the rock sections have shifted sideways, as opposed to up and down (dip-slip).

Structural Map - A map that uses contour lines to show the elevation of underground rock layers.

Structural Trap - A petroleum trap created by the deformation of reservoir rock, such as a fold or fault.

Supply Main - The pipe that receives natural gas at the city gate station and carries it into the distribution system.

Swamp Gas - Almost pure methane that is formed by biological processes (bacterial action) at relatively low temperatures and depths. Also called *microbial gas, marsh gas,* or *biogenic gas.*

Sweetening - Gas conditioning processes that remove carbon dioxide and hydrogen sulfide impurities.

Syncline - A large, long fold of sedimentary rocks that are bent downward.

Tail Gas - The gas stream exiting a natural gas processing plant after the separation of natural gas liquids.

Take or Pay Provisions - Pipeline contract provisions, common during the 1970s-1980s, that required companies to pay producers for the full contractual quantities of gas even if they couldn't sell it.

Tension-leg Platform - A floating offshore wellhead and production platform held in place by large weights on the seafloor, connected by small-diameter, hollow steel tubes.

Thermogenic Gas - Natural gas formed by heat on organic matter underground or by the thermal breakdown of crude oil.

Three-dimensional (3-D) Seismic - A seismic record that shows reflections in three dimensions. These techniques are similar to CAT scans and MRI for obtaining images inside the human body.

Thrust Fault - Faults caused by compression, where one piece of rock has been thrust up and hangs over the lower piece of rock. Thrust faults appear on the surface as mountain ranges.

Tight - Low permeability, resisting fluid flow.

Tight Sands - Gas reservoirs composed of low-permeability sandstone.

Tool Pusher - A drilling company employee at the well site who is in charge of the drilling crews and the rig.

Topographic Map - A map that uses contour lines to show the elevation of the surface of the ground.

Town gas - Combustible fuel made by heating coal. Usually referred to as manufactured gas.

Trader - See Gas Trader.

Trap - An area in reservoir rock where gas and/or oil can accumulate. The trap is overlain by a cap rock.

Trend - An area, or fairway, along which a gas play has been proven and where more fields could be found.

Tubing - Small-diameter steel tube that conducts fluids up the well.

Turbine - Rotary machine used to generate electricity. Turbines can be powered by natural gas, steam, or other fuels.

Turbodrilling - A technique where the drill bit is rotated by a downhole turbine, which is powered by the circulating mud. Because the rotary motion is imparted only at the bit, rotation of the drillstring is unnecessary.

Unbundling - Separation of a pipeline company's purchase services from its transportation services.

Vacuum Furnace - A type of heat treating furnace that creates a vacuum inside to protect the metal product from oxidation.

Vitrinite Reflectance - A geochemical exploration method to determine the maturity of a source rock by examining its content of vitrinite, a type of plant organic matter.

Volatility - The degree to which a financial market experiences fluctuations in price. High volatility means very wide price fluctuations.

Well - A hole drilled into the earth's surface to produce natural gas. Wells can be drilled on land or offshore below the water.

Wet Gas - Natural gas that produces a liquid condensate (heavier hydrocarbons) on the surface, even though it exists as a gas underground.

Wildcat Well - An exploratory well drilled to discover a new gas field.

Wireline Log - Measurements taken by lowering a recording device into the borehole on a cable, or wireline. Wireline logs can record a great variety of physical characteristics including the electrical, radioactive, and sonic properties of the formation rocks and their fluids.

Working Gas - Natural gas in a storage reservoir that can be produced and delivered when needed, as opposed to cushion gas that must remain in place to maintain reservoir pressure.

INDEX

A

Atlas, 21
Auer, Karl, 7
Avoided cost, 128

B

Back offices, 128
Base load, 58, 128
Basis trading, 112-113, 128
Batch, 128
Bi-fuel vehicle, 91-92, 128
Biogenic gas, 3-4, 128
Bit, 128
Blowout, 129
Boiler, 73-74, 86, 129
Booster water heater, 129
Bright spot, 129
Build angle, 129
Bunsen burner, 7
Burner-tip price, 129
Butane, 1-2, 40, 42, 129
By-pass, 129
Byproduct, 1-2, 42

C

Cable-tool drilling, 30, 129
Caliper log, 37, 129
Canadian Gas Association, 107
Canadian Gas Research Institute, 107
Cap rock, 4-5, 129
Carbon dioxide, 1-2, 40-42, 102, 129
Carbonate rock, 18, 129
Casing, 130
Cathodic protection systems, 130
Cavern storage, 57
Cement, 37-38, 130

D

G

H

I

J

K

L

Leaching, 139
Leak control, 67
Leak detector/detection, 53, 139
Leak surveys, 67
Lebon, Philippe, 6
Lighting, 6-7, 94-95
Limestone, 139
Linepack, 139
Liquefied natural gas, 57, 64-65, 139
Liquefied petroleum gas, 1, 57, 64, 139
Liquid products, xvii
Liquid redox process, 139
Liquidity, 139
Lithification, 2, 139
Lithofacies map, 139
Lithographic log, 139
Local distribution company, 61, 113-114
Logging while drilling, 37
Logs, 140

M

Magnetic resonance imaging, 23
Magnetometer, 140
Maintenance, 68-69
Manufactured gas, 5-12, 140
Mapping, 19-21
Marginal reserves, 26
Market basis differential, 140
Market out provisions, 99, 140
Marketers, 111-113, 140
    spot market, 112
    basis trading, 112-113
    futures and options, 113

N

P

R

S

T

U

V

Vacuum furnace, 148
Ventilation air, 80-81
Ventilation standards, 80
Vitrinite, 149
Volatility, 149

W

Water content, 1, 40
Water heater, 9, 74, 76-77, 87, 129
Welding, 46
Well, 149
Well completion, 36-39
Well evaluation, 36-37
Well logging, 36-37
Well stimulation, 40
Well testing, 39-40
Wet gas, 41-42, 149
Wildcat well, 149
Winsor, Frederick, 6
Wireline log, 36-37, 149
Working gas, 149

Z

Zoar field, 56